Ravena Glicéria Noll Diniz
Carlos Fernando Lemos

MANAGEMENT OF SOLID CONSTRUCTION AND DEMOLITION WASTE

Ravena Glicéria Noll Diniz
Carlos Fernando Lemos

MANAGEMENT OF SOLID CONSTRUCTION AND DEMOLITION WASTE

CITY OF BETIM- MG

ScienciaScripts

Cover image: www.ingimage.com

This book is a translation from the original published under ISBN 978-620-6-76089-4.

Publisher:
Sciencia Scripts
is a trademark of
Dodo Books Indian Ocean Ltd. and OmniScriptum S.R.L publishing group

120 High Road, East Finchley, London, N2 9ED, United Kingdom
Str. Armeneasca 28/1, office 1, Chisinau MD-2012, Republic of Moldova, Europe
Printed at: see last page
ISBN: 978-620-7-89806-0

"The future will depend on what we do in the present."

(Mahatma Gandhi)

SUMMARY

Despite the environmental policies established at Eco 92, held in Rio de Janeiro in 1992, there has been little progress since then. The construction sector, although fundamental to the national economy, is recognized as a major generator of environmental impacts due to the unbridled use of natural resources and the significant production of waste. The research highlights the recycling plant in the city of Betim (MG), run by ECOS, as an example of sustainability, converting large volumes of rubble into useful raw materials for the city. The study adopted an applied, descriptive and exploratory research approach, combining quantitative and qualitative methods, through a case study carried out at the Solid Construction Waste Recycling Plant in the Municipality of Betim, located in the metropolitan region of Belo Horizonte, Minas Gerais. The research was organized in the form of articles, addressing different aspects related to CDW. The chapters include a systematic review of alternatives for reducing the environmental impacts of CDW, an analysis of construction waste management in Betim, an evaluation of the current scenario of the recycling plant from a financial, environmental and social perspective, the theoretical potential for generating recyclables in the selective collection of CDW and finally a technical manual for managing public CDW recycling plants. As a result, the research showed different alternatives for waste treatment, as well as demonstrating waste management in the city of Betim, and the results of the last five years, where the plant recycled approximately 150,000 tons of rubble, transforming the recycled aggregate into rural road cover. The plant did not prove to be financially viable under the current management model, registering a negative monthly balance of R$ 1,948,486.89. However, when considering the environmental and social aspects, its relevance to society and the environment is clear. The research also highlighted the importance of implementing the Theoretical Recyclables Generation Plan (PTGR) in parallel with the PDCA cycle to improve waste management in Betim, MG. Finally, it produced a technical management manual based on the *Project Management Body of Knowledge* (PMBoK) that will contribute to project management guidelines for improving processes.

Keywords: Recycling. Sustainability. Management. Plant.

SUMMARY

1. INTRODUCTION

Discussions on sustainability have intensified in scientific circles and in organizations. But in general terms, how effective have these debates been in dealing with solid waste, especially construction and demolition waste?

In 1992, the United Nations Conference on the Environment and Sustainable Development, also known as the Earth Summit, Eco-92 or Rio-92, was held in Rio de Janeiro, Brazil. On that occasion, environmental responsibility policies were established on an agenda that has made little progress in the last 30 years (BRASIL, 1995).

The civil construction sector is one of the sectors that has made the biggest contribution to the national GDP . However, according to Nunes et al. (2020), it has been identified as one of the main causes of environmental impacts. This is due both to the significant consumption of non-renewable natural resources and to the considerable generation of waste, the destination of which is not always properly managed.

According to Zanett *et al.* (2015) the improper disposal of Construction and Demolition Waste (CDW) causes several environmental problems. These include the silting up of rivers, the clogging of urban drainage, as well as serving as a pretext for the irregular deposit of other non-inert waste, encouraging the appearance and multiplication of disease vectors.

In 2002, the National Environment Council (CONAMA) approved Resolution 307/2002, which became the main instrument providing guidelines for Brazilian municipalities to carry out proper environmental management of construction waste.

CONAMA 307/2002 classifies construction and demolition waste (CDW) into four classes: A, B, C and D. As an alternative source for reducing the environmental impacts caused by this waste, one of the options for treating class A waste is recycling. In this context, Cardoso et al. (2014) propose that the disposal of CDW should prioritize recycling and reuse, generating income for those responsible for this process.

To this end, some municipalities have already set up public recycling plants with collection points in their regions, following the guidelines of CONAMA 307/2002.

The study scenario was the municipality of Betim, Minas Gerais (MG), located 30 kilometers from Belo Horizonte, in the metropolitan region of Minas Gerais. Betim is one of the fastest-growing municipalities in the southeast and the fifth most populous in Brazil, with approximately 450,000 inhabitants (IBGE, 2020). The city has one of the highest growth rates in the state of Minas Gerais, and has significant solid waste figures. According to the Betim Municipal Integrated Solid Waste Management Plan (PMGIRS), a total of 82,152,120 Kg/year was generated in 2014. It is estimated that of the total waste generated in the municipality between 1997 and 2014, 23% is solid waste from construction, demolition and renovation (PMGIRS, 2015).

The recycling plant, which is the subject of this research, is managed by the Construction, Works, Services, Projects, Transportation and Transit Company (ECOS). Created in 2018, ECOS took on the responsibility of managing municipal public policies that were previously carried out by municipal departments and companies. One of the main reasons for its foundation was to reduce public service bureaucracy and make the provision of these services more efficient (ECOS, 2019).

In 2019, ECOS, at its Betim unit, transformed 33,600 tons of construction and demolition debris into raw material for the maintenance of rural public roads and asphalt paving in more than 30 neighborhoods in the city (ECOS, 2020).

According to the PMGRS (2015), the management of CDW in Betim, as in most Brazilian cities, has not been providing adequate management of Construction and Demolition Waste/Construction Waste, nor has it been following CONAMA Resolution 307 and ABNT standards (BRASIL, 2015).

1.1 Presentation strategies for the dissertation

The research used the applied research method, of a descriptive and exploratory nature, with a quantitative and qualitative approach, by means of a case study.

The study scenario was the municipality of Betim, Minas Gerais (MG), located 30 km from Belo Horizonte, in the metropolitan region. Betim is one of the Southeast's fastest-growing municipalities and the fifth most populous in Brazil, with approximately 450,000 inhabitants (IBGE, 2020).

To achieve the research objective, a systematic literature review was first carried out to provide a reference for the study in question, focusing on alternatives to minimize the impacts caused by CDW.

According to Marconi and Lakatos (2017), the quantitative-descriptive method aims to delineate or analyze the characteristics of facts or phenomena, evaluate programs or isolate main or key variables.

In the systematic review of article 1, a descriptive qualitative bibliographic survey methodology was used. The study included articles from periodicals and annals, as well *as* undergraduate and graduate course conclusion papers.

Articles 2 and 3 used the applied research method, of a descriptive nature, with a qualitative approach, by means of a case study.

Field surveys were carried out, with periodic visits to the ECOS company to collect the data needed for the research and analysis of waste management and the production and recycling process. In addition, a questionnaire was drawn up to analyze the management and implementation of the plant, aimed at those directly involved in the investigation.

In view of this, and as indicated in the introduction, this research was structured in the form of articles. Therefore, in the subsequent subsections, the methodology of each article in this dissertation is presented:

a) **CHAPTER I:** Alternatives for reducing the environmental impact of solid construction and demolition waste - a systematic review.

b) **CHAPTER II:** Case study: Analysis of construction and demolition solid waste (CDSW) management in the municipality of BETIM -MG.

c) **CHAPTER III:** Case study: Study of the current scenario of the solid construction waste recycling plant in the municipality of Betim/mg

d) **CHAPTER IV:** The Theoretical Potential for Generating Recyclables (PTGR) in the selective collection of construction and demolition solid waste (CDSW) in the municipality of Betim/MG

e) **CHAPTER V:** Technical Manual for the Management of Public Construction and Demolition Waste Recycling Plants

2 BACKGROUND

When it comes to sustainable development, it is necessary to seek a balance between economic interests, the preservation of environmental resources and quality of life. According to Cruz and Marques (2014), local governments must promote efficient public management and seek improvements in economic performance.

In order to gain a competitive advantage in the market in a scenario of environmental changes and growing popular awareness, sustainability has been adapting methods with the aim of minimizing human impacts on the environment. This is done by harmonizing economic, social and environmental elements, which together form the sustainability tripod.

The environmental pillar, according to Ribas (2017), refers to an organization's ability to pursue its production processes and offer sustainable services and goods at competitive prices.

In this way, public management plays a fundamental role in balancing efficiency in the management of public and private resources and investments, and in promoting the guarantee of a healthy environment for current and future generations.

According to Cavalcanti and Silveira (2016), project management consists of applying knowledge, skills, tools and techniques to meet the project's objectives. Therefore, applying management tools to the recycling plant increases its chances of success.

According to the *Project Management Institute* (PMI), the *Project Management Body of Knowledge* (PMBoK) encompasses proven traditional methods that are applied to project management in its entirety, as well as innovative practices that are emerging to achieve effective management.

In this context, the research aims to adapt the PMBoK processes to the reality of public services at the recycling plant in Betim, with the aim of making its operation viable. To this end, tools from project and process management and production engineering will be used to estimate the cost of operating the plant and define its economic viability, demonstrating the potential of recycling. The research will also seek to comply with legislation, international agreements and standards designed for this purpose.

In addition, the research generated a technical product called the Technical Manual for the Management of Public Construction and Demolition Waste Recycling Plants, with the aim of collaborating with the efficient public management of the plant in Betim.

The hypotheses addressed to achieve the research objectives were: How solid construction waste is managed in Betim and at the recycling plant; are there any options for reducing the environmental impacts caused by construction waste; what is the plant's current scenario from a financial, environmental and social perspective; what is the theoretical potential of construction waste; what is the possibility of improving the plant's management?

Research and development are fundamental to the sustainable growth of any organization. That said, encouraging the adoption of effective CDW management techniques is an action that reduces the environmental impacts caused by this waste, as well as ensuring efficient management. This ensures that the economic factor is not overlooked and promotes the economic sustainability of the plant in the recycling process.

3 OBJECTIVES

3.1 General Objective

To analyze the management of solid construction and demolition waste in Betim, Minas Gerais, which is sent to a recycling plant.

3.2 Specific objectives

To achieve this goal, the following specific objectives will have to be met:

a) To draw up a systematic review of CDW with a focus on minimizing the impacts caused by this waste.

b) Carry out an analysis of the municipality's CSR management

c) Mapping the plant's production process, defining the starting and end points of activities in all phases using the model VSM - value stream mapping - a *Lean Manufacturing* tool.

d) Carry out an economic feasibility analysis of the plant.

e) Draw up a technical report on the potential for generating construction waste

f) Developing a project management guide based on the PMBoK

4 CHAPTER I

ALTERNATIVES FOR REDUCING THE ENVIRONMENTAL IMPACT OF SOLID CONSTRUCTION AND DEMOLITION WASTE - A SYSTEMATIC REVIEW

4.1 SUMMARY

Despite its importance to society, the construction industry has been responsible for significant environmental impacts in its production process, from the extraction of raw materials to the improper disposal of waste in the environment. The aim of this research was to analyze the environmental impacts of solid construction and demolition waste in Brazil and to identify technologies for reducing and reusing this waste. The method used was a descriptive qualitative approach to a bibliographic survey of national literature published between 2013 and 2023, using the literature review protocol - RSL and Start *software* to help extract the papers. As a result, 18 articles were found which met the selection and quality criteria defined during the systematic literature review protocol, which provided analyses of CSR as well as alternatives for minimizing the impacts caused to the environment. As an academic justification, this study aims to contribute to theoretical discussions on the subject of construction waste management and its impacts on the environment in the scientific environment, as well as proposing alternatives for reducing and encouraging the non-generation of CDW.

Keywords: Construction technology. Recycling. BIM. PBQP-H. CONAMA 307.

4.2 ABSTRACT

Despite its importance to society, the construction industry has been responsible for significant environmental impacts in its production process, from the extraction of raw materials to the improper disposal of waste in the environment. The aim of this research was to analyze the environmental impacts of solid construction and demolition waste in Brazil and to identify technologies for reducing and reusing this waste. The method used was a descriptive qualitative approach to a bibliographic survey of national literature published between 2013 and 2023, using the RSL literature review protocol and Start software to help extract the papers. As a result, 18 articles were found, which correspond to the selection and quality criteria defined during the execution of the systematic literature review protocol, which brought analysis of CSR as well as alternatives to minimize the impacts caused to the environment. As an academic justification, this study aims to contribute to theoretical discussions on the subject of construction waste management and its impacts on the environment in the scientific environment.

Keywords: Technology in civil construction. Recycling. BIM, PBQP-H. CONAMA 307.

4.3 INTRODUCTION

According to Simão et al. (2019), civil construction plays a very important role, involving a range of important activities for the economy, as it contributes substantially to the country's economic and social development, as well as being one of the sectors that most degrades the environment. Due to urban growth in recent decades, the generation of waste from the sector's activities has become a matter of debate in order to reduce the impacts caused by this waste. According to Teixeira, Teixeira and Rocha (2020), the construction sector is one of the largest in the world economy, employing around 7% of the working-age population, with annual spending of more than 10 trillion dollars on goods and services related to the activity.

Despite its economic importance and accelerated urban growth, the construction sector is one of the sectors that most degrades the environment throughout its production chain. Souza (2021) mentions that through accelerated urbanization processes and an increase in the population of cities, the world is faced with debates on how to reverse or mitigate the impacts of the development of urban centers,

According to Hammes, Souza and Rodriguez (2021), the depletion of raw materials has led to a process of environmental awareness in business organizations through a search for products and processes that are considered environmentally friendly.

As a result, the construction sector plays an important role in the search for sustainable practices. Furuya and Oliveira (2018) report that the construction sector is increasingly demanding the adoption of sustainable practices, from the start and planning of a construction project to the inauguration and operation period.

Improper disposal of CDW can lead to serious problems such as soil and water pollution, the development of diseases and threats to life on Earth, especially for future generations.

Inadequate disposal of CDW causes siltation of rivers, clogging of urban drainage, as well as serving as a pretext for the irregular deposit of other non-inert waste, encouraging the appearance and multiplication of disease vectors (ZANETT *et al.,* 2015).

In view of the above, the aim of this review is to explore the subject of solid construction waste and some alternatives for treating and reducing its generation. As an academic justification, this study aims to contribute to theoretical and practical discussions on the subject of construction waste management in the scientific environment. The systematic review methodology was adopted for this study.

4.4 METHODOLOGY

The research used a qualitative, descriptive approach to a bibliographical survey, and a systematic review was used as a technique for this survey. The following materials were considered for this study: journal articles and annals, *undergraduate* and graduate course conclusion papers.

A priori, a protocol was drawn up following the guidelines of the RSL-Systematic Literature Review for better planning of this review. The document was executed in Microsoft word software and then transferred to the START program, which Professor Valdick recommended for organizing the RSL.

All the stages followed a flowchart based on the planning, collection, analysis and results phases. Figure 1 describes the strategies used and designates each one.

Figure 1 - Flowchart of the RSL stages

Source: Prepared by Diniz R.G.N, 2023.

4.5 SEARCH STRATEGY

The search strategy was carried out in order to organize the work and have a detailed view of the research objectives, following the steps described below:

4.5.1 Research questions

The research questions were: "What are the impacts caused by RSCD and what are the alternatives for reducing these impacts?"

This RSL is derived from the definition of the elements shown in Table 1.

Chart 1 - Criteria adopted for the research

Critérios	Descrição
População	Produção cientifica sobre os RSCD
Intervenção	Leitura e análise dos RSCD, seus impactos e alternativa para reduzir esses impactos
Controle	Artigos que contenham análise dos RSCD, gestão desses e elencam alternativas para redução dos impactos ambientais
Resultado	Alternativas para minimizar os impactos causados pelos RSCD.
Contexto de Aplicação	Com o crescimento do setor da construção civil, faz-se necessário estudos que tragam alternativas para reduzir ou mitificar os impactos causados pelos RSCD.

Source: Prepared by Diniz R.G.N, 2023.

Table 2 presents the research question(s) for this RSL:

Chart 2 - Research questions

Pergunta	Descrição da Pergunta
P1	Quais impactos os RSCD causam ao meio ambiente?
P2	Como realizar a gestão adequada desses resíduos?
P3	Qual alternativa para reduzir esses impactos?

Source: Prepared by Diniz R.G.N, 2023.

4.5.2 Databases and research methods

Table 3 below describes the databases and methods used in the research .

Quadro 3 - Descrição das bases de dados e dos métodos utilizados na pesquisa

Métodos de pesquisa de fontes	Serão consideradas fontes disponíveis via WEB. exclusivamente em bases de dados científicas, a pesquisa se dará através de busca de palavras chaves pesquisadas em portais de busca de artigos, teses, dissertação e periódicos. Serão consideradas as Strings encontradas em títulos, resumos e palavras chaves nas bases de dados pesquisadas. Após análise dos resumos e verificando a coerência, metodologia e objetivo, este, será selecionado para leitura em sua totalidade, e após será aceito ou rejeitado conforme os critérios de inclusão e exclusão definidos nessa pesquisa.
Bases a serem pesquisadas	Scopus; Web of Science; Scielo; doaj directory of open access journals; Scirus; Science Direct; Capes Periódcos; Google Academic;
Idiomas	Português e inglês (desde que seja páginas do Brasil
String de busca	Resíduos Sólidos da construção e demolição Impactos dos RSC Usinas de reciclagem dos resíduos da construção Agregados reciclados da construção e demolição Tecnologias aplicadas a gestão de resíduos da construção civil
Definição de tipos de estudos	Este estudo define os tipos de estudos primários que serão selecionados durante a execução da revisão sistemática com abordagem quantitativa descritiva. Foi elaborado a partir de uma revisão de literatura nacional disponibilizadas em periódicos. Tal levantamento fio feito seguindo os protocolos do Software START.

Source: Prepared by Diniz R.G.N, 2023.

Table 4 describes the search *strings* adapted for each database searched

Table 4 - Search *strings*

Base de dados	String Adaptada/Utilizada	Filtros e tipo de busca
Scopus	Resíduos Sólidos da construção e demolição Usinas de reciclagem dos resíduos da construção Agregados reciclados da construção e demolição Tecnologia aplicadas a gestão de resíduos da construção civil	Não foram aplicados filtros, somente utilizados o campo buscar aplicando a string.
Web of Science	Resíduos Sólidos da construção e demolição Usinas de reciclagem dos resíduos da construção Agregados reciclados da construção e demolição Tecnologia aplicadas a gestão de resíduos da construção civil	Não foram aplicados filtros, somente utilizados o campo buscar aplicando a string.
Scielo	Resíduos Sólidos da construção e demolição Usinas de reciclagem dos resíduos da construção Agregados reciclados da construção e demolição Tecnologia aplicadas a gestão de resíduos da construção civil	Não foram aplicados filtros, somente utilizados o campo buscar aplicando a string.
doaj directory of open access journals	Resíduos Sólidos da construção e demolição Usinas de reciclagem dos resíduos da construção Agregados reciclados da construção e demolição Tecnologia aplicadas a gestão de resíduos da construção civil	Não foram aplicados filtros, somente utilizados o campo buscar aplicando a stringS.
Capes Periódcos:	Resíduos Sólidos da construção e demolição Impactos ambientais do Resíduos Sólidos da construção e demolição Usinas de reciclagem dos resíduos da construção Agregados reciclados da construção e demolição Tecnologia aplicadas a gestão de resíduos da construção civil	Foi utilizado o campo buscar aplicando a strings. Foram aplicados filtros ano de 2013 a 2023, Idioma Inglês e Português e coleção das bases de Web of Science; Scielo; doaj directory of open access journals
Google Acadêmico:	Resíduos Sólidos da construção e demolição Impactos ambientais do Resíduos Sólidos da construção e demolição Usinas de reciclagem dos resíduos da construção Agregados reciclados da construção e demolição Tecnologia aplicadas a gestão de resíduos da construção civil	Foi utilizado o campo buscar aplicando a strings, foram aplicados filtros ano de 2013 a 2023, Idioma Inglês e Português

Source: Prepared by Diniz R.G.N, 2023.

4.5.3 Inclusion criteria

The inclusion criteria for the studies being researched are shown in Table 5:

Q Table 5 - Inclusion criteria

Critério	Descrição do Critério de Inclusão
CI1	Serão incluídos trabalhos que realizaram análise dos RSCD.
CI2	Serão incluídos trabalhos publicados e disponíveis integralmente nas bases científicas buscadas
CI3	Serão incluídos trabalhos que tragam alternativas para reduzir os impactos ambientais dos RSCD
CI4	Serão incluídos trabalhos escritos no período de 2013 a 2023
CI5	Serão incluídos trabalhos sobre reciclados agregado
CI5	Automatically classified by SCAS

Source: Prepared by Diniz R.G.N, 2023

4.5.4 Exclusion criteria

The Exclusion Criteria for the studies being researched are shown in Table 6:

Chart 5 - Inclusion criteria

Critério	Descrição do Critério de Exclusão
CE1	Serão excluídos trabalhos que não apresentem resumo/abstract
CE2	Serão excluídos trabalhos que não estejam nas linguagens português ou inglês
CE3	Serão excluídos trabalhos que não foram escritos nos últimos 10 anos
CE4	Serão excluídos trabalhos que não tenham coerência com o tema
CE5	Serão excluídos trabalhos que não contenham as palavras chaves em seu resumo
CE6	Serão excluídos trabalho que não sejam do Brasil
CE7	Automatically classified by SCAS

Source: Prepared by Diniz R.G.N, 2023

4.5.5 Quality criteria

The Quality Criteria of the retrieved studies are shown in Table 7:

Table 7- Quality criteria

Critério	Descrição do Critério de Qualidade
CQ1	Os métodos aplicados foram descritos de forma objetiva?
CQ2	O texto foi escrito com coerência?
CQ3	Houve análise de gestão dos resíduos sólidos?
CQ4	Explicitou técnicas para redução ou reaproveitamento dos RSCD?

Source: Prepared by Diniz R.G.N, 2023

4.5.6 Quality form fields

Quality criteria adopted in this RSL presented in Table 8

Table 8 - Quality criteria fields

Coerência do texto	Sim, Não
Coesão textual	Sim, Não
A análise dos RSCD foi feita de forma objetiva	Sim, Não
Elencou os impactos ambientais causados pelos RSCD	Sim, Não
Trouxe opções de tecnologia para reduzir ou reaproveitar os RSCD	Sim, Não

Source: Prepared by Diniz R.G.N, 2023

4.5.7 Information extraction

Once the primary selection of articles has been made, the extraction of information relevant to the proposed topic begins.

In this section, the criteria and results were analyzed by reading the abstract and conclusion of the papers. The extraction criteria fields are shown in Table 9:

Table 9 - Extraction criteria fields

Campo	Tipo	Conteúdo
Metodologia	Text	
Objetivo coerente com o tema	Pick on list	{Sim,Não}
Análise dos Resultados	Text	
Houve análise de impactos dos RSCD	Pick on list	{Sim,Não}
Trouxe de forma coerente as alternativas par redução ou reutilização dos RSCD	Pick on list	{Sim,Não}
Elencou leis ou normativas de tratamentos do RSCD	Pick on list	{Sim,Não}
Houve análise de dados	Pick on list	{Sim,Não}

Source: Prepared by Diniz R.G.N, 2023.

4.6 RESULTS AND DISCUSSION

4.6.1 Publication Selection Process

Firstly, a protocol was drawn up to formulate the objectives of the systematic review, and within the protocol, search *strings* were created with the theme CSDR.

After searching each database, the files were exported in BIBtex mode to a specific folder assigned to each database. After saving each file in its respective folder, they were exported to the *Start* tool to help with the qualification and extraction process.

Considering the search *strings*, a total of 839 papers were found in the databases cited in the protocol.

In the selection stage, the titles of each study were first analyzed. As the *strings included* the words environmental impacts, technology, waste, plants, recycling and civil construction, 634 papers were rejected because they dealt with topics that would not be included in this review and did not meet the selection criteria described in the protocol, such as excluding papers written more than 10 years ago, papers that were not national or papers that did not include keywords from the *strings*.

In addition to the articles mentioned in the paragraph above, 95 articles were duplicates*, Automatically Classified* was also considered, so 729 articles were excluded in the first qualification phase and did not go on to the subsequent extraction phase. For the extraction phase, 110 papers were considered.

In this phase, the abstracts and conclusions were read, a stage called the extraction phase, where there was a refinement in the selection of articles by analyzing the quality, exclusion and inclusion criteria again, the *Start software was* used to help, in this way, the selection and quality criteria were done manually, answering the criteria, creating Score numbers in the *software* itself.

Those that met the greatest number of criteria were evaluated. This report is created using *Start*, which generates an Excel file for better visualization of the results.

To do this, we separated the papers with the highest number of scores and arranged the authors' arguments in these papers. In addition to carefully evaluating the quality of the papers, after the analysis we had 87 rejected papers and 23 accepted papers.

The next stage required a complete reading of the articles to assess the quality of the studies and extract the information at this stage. It was possible to observe papers written with the same purpose and similar approaches. For this reason,

6 were discarded, and it was decided to work with 17 articles chosen to make up the final construction of this investigation, as shown in Table 10 below.

Table 10: Selected final articles

ID Paper	Title	Authors	Status/Selection
1	Sustentabilidade ambiental urbana: análise da influência da vegetação em parâmetros ambientais.	Meiry Helen Sousa Bordim - Regina Marcia Longo - Bruno Sousa Bordim	Accepted
2	Análise bibliométrica das publicações em quatro eventos científicos sobre gestão de resíduos sólidos urbanos a partir da Política Nacional de resíduos Sólidos – Lei nº 12.305/2010	Estela Cristina Bonjardim -Raquel da Silva Pereira - Eliana Vileide Guardabassio	Accepted
3	Análise comparativa do levantamento de quantitativos entre o método manual e a plataforma bim	Renan Caminha Marinho	Accepted
4	Utilização do método fuzzy topsis para avaliação de alternativas de reciclagem dos resíduos sólidos da construção civil: utilização do método fuzzy topsis para avaliação de alternativas de reciclagem dos resíduos sólidos da construção civil	Deoclécio Júnior Cardoso - Diogo Daniel Görgen Kogler - Julia Tontini - Antonio Vanderlei dos Santos- Julio Cezar Mairesse Silulo	Accepted
5	O reconhecimento das certificações ambientais no setor da construção civil	Danielle Elis Garcia Furuya - Michelle Taís Garcia Furuya - Rodrigo Coladello De Oliveira	Accepted
6	POLÍTICA NACIONAL DE RESÍDUOS SÓLIDOS: perspectivas de cumprimento da Lei 12.305/2010 nos municípios brasileiros, municípios paulistas e municípios da região do ABC	Maria Helena Scalabrin Cardoso Gomes - Edenis Cesar Oliveira - Luís Paulo Bresciani - Raquel da Silva Pereira	Accepted
7	A gestão de resíduos da construção civil no município de Sorocaba-SP	Daniel Bertoli Gonçalves	Accepted
8	Avaliação da geração de resíduos de construção civil (RCC) no processo de assentamento cerâmico	L. P. GUIMARÃES1, M. GIANEZINI2, V. M. BRISTOT3, Y. B. EMERIM4 e M. L. F. GUIMARÃES	Accepted
9	Avaliação de desempenho da logística reversa no setor da construção civil	Gabriela Hammes - Eduarda Dutra de Souza - Carlos Manuel Taboada Rodriguez	Accepted
10	Uso de indicadores na gestão dos resíduos sólidos urbanos: uma proposta metodológica de construção e análise para municípios e regiões	Suellen Silva Pereira - Rosires Catão Curi - Wilson Fadlo Curi	Accepted
11	Serviços de resíduos sólidos à luz da política pública de universalização da Lei n. 12.305/ 2010: um estudo de caso	Christopher Abreu Ravagnani - José Carlos de Oliveira	Accepted
12	Percepção Socioambiental dos Moradores sobre a Disposição Inadequada de Resíduos da Construção Civil (RCC) de um Bairro em Expansão na cidade de Sobral/CE	Rennan Liberato Rodrigues - Ana Lúcia Feitoza Freire Pereira - Francisco Amílcar Moreira Junior	Accepted
13	Análise dos impactos ambientais causados pelos resíduos sólidos de construção e demolição em conceição do almeida – BA	Izáira Cunha Santana	Accepted
14	A Densidade e Evolução Urbana como Ferramentas de Planejamento Urbano Sustentável: O Caso em Primavera do Leste - MT, Brasil	Maria Lívia Resende Souza - Gustavo Macedo de Mello Baptista	Accepted
15	Estudo Prospectivo Sobre Inteligência Artificial Aplicado ao Setor da Construção Civil	Fabio dos Santos Teixeira - Paulo dos Santos Teixeira - Carlos Alberto Machado da Rocha	Accepted
16	Resíduos sólidos da construção civil	Gabriela Candeo Zanetti - Karen Staquecini Martines - Maria Eunice Carvalho Tosello - Rebeca Delatore Simões	Accepted
17	Análise comparativa do levantamento de quantitativos entre o método manual e a plataforma bim	RENAN CAMINHA MARINHO	Accepted

Source: Prepared by Diniz R.G.N, 2023.

In their studies, Bordim, Longo and Bordim (2022) reported that with the start of the industrialization process, general structural changes took place, forming new municipalities and starting urbanized areas, which accelerated environmental degradation. In this context, a variety of critical thoughts have been observed regarding the acceleration of the industrial process and the acceleration of urban spaces, causing major environmental impacts.

Despite this critical thinking, some authors, such as Bonjardim, Pereira and Guardabassio (2018), argue that the continuity of the construction sector to this day has been fundamental for socio-economic development, although it has intensified pollution and environmental devastation.

The construction sector is one of the sectors that has made the biggest contribution to the national GDP. However, according to Gonçalves (2016), despite its undeniable importance, it has been identified as one of the biggest generators of environmental impacts, whether due to the significant consumption of non-renewable natural resources or the intense generation of waste, which is not always properly disposed of.

Santana (2016), in her final paper, mentions that the negative environmental impacts associated with CDW are due to the irregular disposal of this waste. According to the author, the most common impacts of this waste are: attracting vectors, obstructing streets and surface drainage; visual pollution and attracting other types of waste. This makes it necessary to look for alternatives to minimize the impacts caused by this waste.

Accordingly, Brazil has created laws to regulate the treatment of solid waste and solid waste from construction and demolition.

Ravagnan and Oliveira (2018):

> "The National Solid Waste Policy (PNRS), Law n. 12.305/2010 - the regulatory framework for solid waste services in Brazil - regulated by Decree n. 7.404/2010, consolidated principles, objectives and instruments, as well as guidelines for integrated management with a view to the environmentally appropriate management of solid waste."

Gomes *et al.* (2013) reports that the " PNRS - National Solid Waste Policy , determined that Brazilian municipalities should draw up and submit their respective Solid Waste Management Plans by 2012, imposing the eradication of so-called "dumps" by the year 2014.

In 2002, the National Environment Council (CONAMA) enacted Resolution 307/2002, which, according to Gonçalves et al. (2012), "is today the main instrument guiding Brazilian municipalities so that they can effectively contribute to the proper environmental management of construction waste".

CONAMA 307/2002 classifies solid construction and demolition waste into four categories: Class A waste: waste that can be reused or recycled as aggregates; Class B waste: waste that can be recycled for other uses, such as plastics, paper, cardboard, metals, glass, wood and plaster; Class C waste: waste for which no economically viable technologies or applications have been developed to enable it to be recycled or recovered; Class D waste: hazardous waste from construction, such as paints, solvents, oils and others, or waste that is contaminated or harmful to health.

Considering the authors' ideas for the treatment of CDW, one of the alternatives for the treatment of this waste, and the most discussed among them, is recycling. In this context, Da Silva *et al.* (2019) state that the destination to be given to CDW should prioritize recycling and reuse, generating income for those responsible for this process and avoiding the occupation of space by raw materials extracted from nature.

A second alternative for mitigating the environmental impacts caused by this waste would be to reuse it on site. In this context, Zanettil *et al.* (2015) discusses the use of waste, one or more times, in its original form, without processing, i.e. without physical or physico-chemical transformation.

Another option for solving the problems generated by this waste would be to create environmental indicators, according to Pereira, Curi and Curi (2016):

> "Indicators organized in such a way as to support decision-making, through diagnoses and forecasts, used to promote specific policies, such as MSW management, prove to be an indispensable way of seeking urban environmental sustainability."

Technological advances are important in helping to solve the problem of construction waste. Several authors have raised this issue, pointing out that during the stages of a construction project, materials are wasted. This waste can have various causes, such as poorly designed projects, inadequate execution of the work or failures in materials management.

In 1998, the Ministry of Cities created the Programa Brasileiro da Qualidade e Produtividade do Habitat (PBQP-H), based on ISO 9000, a program to modernize construction work.

Guimarães *et al.* (2019) defines that the objective of the PBQPH is to organize the construction sector, improving the quality of the *habitat* and generating productive modernization.

One alternative to avoid wasting materials and have a sustainable construction is the technology *Building Information Modeling (*BIM), created to minimize the impacts caused to the environment by the construction sector according to Catelani (2016).

For the author, the tool is a technological platform applied to civil construction that makes it possible to model the projects and details of the specifications of the installations or the building, avoiding waste and costs in construction (CATELANI, 2016).

Marinho (2017) proposes that BIM is a set of policies, processes and technologies that, together, are capable of managing, coordinating and planning the design process of a building or facility, as well as carrying out prior performance tests, managing and storing information and data, throughout the entire life cycle of the work, from conception to conceptualization of the idea, project development, construction and post-construction.

That said, the research has shown some methods of dealing with solid waste generated by the construction sector in order to avoid environmental degradation caused by the spread of construction work.

However, Rodrigues, Pereira and Moreira Junior (2017) report that:

> "The problem is more intense in medium-sized and small construction projects, especially those related to renovations and demolitions, above all due to the lack of efficient management by inspection bodies, as well as the lack of an environmental culture and awareness on the part of builders and the population in general."

Therefore, more studies are needed, including new technologies to cater for this contingent, as well as greater monitoring and promotion of conscious environmental practices.

4.7 CONCLUSION

The research did what it set out to do by exploring the subject of Solid Construction Waste, using the approach of various authors and demonstrating alternatives for managing and dealing with this waste.

Complying with legislation by drawing up the Municipal Integrated Solid Waste Management Plan is the first step towards effective management of solid waste and the way to ensure compliance with the 2030 agenda, as demonstrated in the research.

In order to avoid environmental degradation due to the incorrect disposal of CDW, the authors cited in the research reaffirm the importance of encouraging recycling, creating new alternatives that reduce the generation of this waste, meeting consumption needs without compromising the environment and future generations, complying with legislation and fulfilling the proposals of Brazil's 2030 agenda.

4.8 REFERENCES

BONJARDIM, Estela; PEREIRA, Raquel; GUARDABASSIO, Eliana. **Bibliometric analysis of publications in four scientific events on urban solid waste management based on the national solid waste policy - law no. 12.305/2010**. Desenvolv. Meio Ambiente, São Caetano do Sul, SP, Brazil, year 2018, v. 46, n. 10.5380, p. 313-336, May 11, 2018.

BORDIM, Meiry Helen; LONGO, Regina; BORDIM, Bruno. **Urban environmental sustainability:** analysis of the influence of vegetation on environmental parameters. **Revista de Gestão Ambiental e Sustentabilidade, Campinas** - SP, ano 2022, n. e19447, p. 1-24, 3 fev. 2022.

CARVALHO, P.A.P.; MOREIRA, L.G.S.L.; Barbosa, BB. **Reusing solid waste to make concrete blocks**. Teccen Magazine. 2019 Jul/Dec; 12 (2): 102-108.

CATELANI, Wilt Silva (Brazil). CBIC - **Brazilian Chamber of the Construction Industry. BIM implementation collection for construction companies and developers: BIM fundamentals** - v.1'. Brasilia: Gadioli Cipolla Branding e Comunicação, 2016. 117 p. Available at:. Accessed on: April 12, 2017.

DA SILVA, Deoclécio Júnior Cardoso *et al.* **Utilização** do **método fuzzy topsis para avaliação de alternativas de reciclagem dos resíduos sólidos** da **construção civil** , Santo Ângelo |, ano 2019, v. 8, n. 1, p. 100-117, 28 dez. 2019.v Federal do Ceará, Fortaleza, 2017.

FURUYA, Danielle; FURUYA, Michelle; OLIVEIRA, Rodrigo. **The recognition of environmental certifications in the construction sector**. Colloquium :Exactarum, [*s. l.*], year 2018, v. 10, n. 10.5747, p. 203, 2018.

GOMES, Maria Helena; OLIVEIRA, Edenis; BRESCIANI, Luís; PEREIRA, Raquel. **National solid waste policy: prospects for compliance with law 12.305/2010 in Brazilian municipalities, São Paulo municipalities and municipalities in the abc region**. Rev. Adm. UFSM, Santa Maria - BR, ano 2014, v. 7, n. 10.5902, ed. 1983465913026, p. 93-110, 4 mar. 2013.

GONÇALVES, D. B. **The management of construction waste in the municipality of sorocaba-sp**. REEC - Electronic Journal of Civil Engineering, Goiânia, v. 11, n. 2, 2016. DOI: 10.5216/reec. V11i2.35791. Available at: https://revistas.ufg.br/reec/article/view/35791. Accessed on: September 19, 2023.

GUIMARÃES, L. P.; GIANEZIN, M.; BRISTOT, V. M.; EMERIM, Y. B.; GUIMARÃES, M. L. F. **Evaluation of the generation of construction waste (rcc) in the ceramic laying process.** HOLOS, [*s. l.*], year 35, v. 3, ed. e6694, 2019.

HAMMES, Gabriela; SOUZA, Eduarda; RODRIGUEZ, Carlos Manuel T. **Performance evaluation of reverse logistics in the construction sector**. Exacta : Production Engineering, [*s. l.*], year 2021, n. 497-522, p. 1,27, July 12, 2021.

MARINHO, RENAN CAMINHA. **Comparative analysis of quantity surveying between the manual method and the BIM platform**. Federal University of Ceará, Technology Center, Civil Engineering Course. 2017. 97 f. Course completion work (Graduation) - Federal University of Ceará, Technology Center, Civil Engineering Course, [*S. l.*], 2017.

PEREIRA, Suellen Silva; CURI, Rosires Catão; CURI, Wilson Fadlo. **Use of indicators in urban solid waste management: a methodological proposal for construction and analysis for municipalities and regions**. Eng. Sanit. Ambient, Campina Grande -PB, ano 2018, v. 23, n. 3, p. 471-483, 3 maio 2016.

RAVAGNAN, Christopher; OLIVEIRA, José Carlos. **Solid waste services in the light of the public universalization policy of LAW No. 12.305/2010**: a case study. Revista Direito Ambiental e sociedade, [*s. l.*], year 2018, v. 8, n. 3, p. 245-267, September 2018.

RODRIGUES, Rennan Liberato; PEREIRA, Ana Lúcia Feitoza Freire; MOREIRA JUNIOR, Francisco Amílcar. **Socio-environmental perception of residents on the inadequate disposal of construction waste (RCC) in an expanding neighborhood in the city of Sobral/CE**. Boletim do Observatório Ambiental Alberto Ribeiro Lamego, Campos dos Goytacazes/RJ, year 2017, v. 1, n. 2, p. 7-18, 2017.

SANTANA, IZÁIRA CUNHA. **Analysis of the environmental impacts caused by solid construction and demolition waste in conceição do almeida** - BA. Federal University of the Recôncavo da Bahia, 2016. 58 p. v. 2.

SIMAO, A. dos S., ALCOFORADO, L. F., LONGO, O. C., SANTOS, D. A. dos, SANTOS, F. dos, SILVA, A. D., Menezes, C. A. G., & Junior, J. C. M. (2019). **Impacts of industry 4.0 on brazilian civil construction. Brazilian** JournalofDevelopment, 5(10), 19670-19685. https://doi.org/10.34117/bjdv5n10-183

SOUZA, Maria. **Density and urban evolution as tools for sustainable urban planning: the case of primavera do leste - MT, Brazil. Cadernos de arquitetura e urbanismo: Paranoá.** Brasília, Brazil, year 30.2021. DensityandUrban Evolution as a SustainableUrban Planning Tools: the Case in Primavera do Leste - MT, Brazil - Institute of Geosciences, University of Brasília, Brasília, Brazil. gmbaptista@gmail.com ORCID ID: 0000-0002-1973-2725 Received on 24/10/2020. Accepted 29/03/2021, p. 1-20, 29 Mar. 2021.

TEIXEIRA, Fabio; TEIXEIRA, Paulo; ROCHA, Carlos. **Prospective study on artificial intelligence applied to the construction sector.** Federal Institute of Education, Science and Technology of Pará, Belém, PA, Brazil, year 2020, n. 10.9771, p. 1-13, Feb. 1, 2020.

ZANETTI1, Gabriela Candeo; MARTINES, Karen Staquecini; TOSELLO, Maria Eunice Carvalho Tosello; SIMÕES, Rebeca Delatore. **Solid construction waste**. Encontro Nacional de Ensino, Pesquisa e Extensão, Presidente Prudente, ano 2015, v. 12, ed. especial, p. 350-358, 22 oct. 2015.

5 CHAPTER II

CASE STUDY: ANALYSIS OF THE MANAGEMENT OF SOLID CONSTRUCTION AND DEMOLITION WASTE (SDW) IN THE MUNICIPALITY OF BETIM - MG

5.1 SUMMARY

Eco-92 established environmental responsibility policies, on an agenda that has made little progress in the last thirty years. It is therefore necessary to step up our efforts to collect, analyze and discuss information on the management of solid construction and demolition waste (SSCD). The aim of this article was to diagnose the management of CDW in Betim. The Brazilian Association of Public Cleaning and Special Waste Companies (ABRELPE) highlights the Southeast region as the largest generator of solid construction and demolition waste in Brazil, representing approximately 50 percent of the total generated in the country. This work used the applied research method, of a descriptive nature, from a qualitative approach, by means of a case study. The data was obtained from internal production control documents of the ECOS company and from the Integrated Solid Waste Management Plan of the Municipality of Betim - Minas Gerais (MG), for the period 2015 to 2022. The data on waste generation in 2015 and subsequent generation and recycling since the operation of the Betim Recycling Plant in 2018 to 2022 were analyzed. The results of this research show that the municipality proposed the Municipal Integrated Solid Waste Management Plan in 2015, in accordance with current legislation, with targets that have been met since 2018. Thus, in the last five years, it has recycled approximately 150,000 tons, transforming the recycled aggregate into rural road cover, enforcing compliance with the 2030 agenda, preserving the ecosystem and becoming a model city for other municipalities.

Keywords: Sustainability. Agenda 2030. Environmental responsibility. Municipal Integrated Solid Waste Management Plan.

5.2 ABSTRACT

Eco-92 established policies for environmental responsibility, in an agenda that has advanced little in the last thirty years. Therefore, it is necessary to intensify our efforts to collect, analyze and discuss information about the management of solid construction and demolition waste (CSWR). This article aimed to diagnose the management of CSWR, in Betim. The Brazilian Association of Companies for Public Cleaning and Special Waste highlights the Southeast region as the largest generator of solid waste from construction and demolition in Brazil, representing approximately 50 percent of the total generated in the country. This work used the method of applied research, of a descriptive nature, from a qualitative approach, through a case study. The data was obtained through internal production control documents of the ECOS Company and the Integrated Solid Waste Management Plan of the Municipality of Betim (MG), for the period 2015 to 2022. The data of waste generation in 2015 and subsequent generation and recycling since the operation of the Betim Recycling Plant in 2018 to 2022 were analyzed. They found as results of this research that the municipality proposed the Municipal Plan of Integrated Management of Municipal Solid Waste, in 2015, according to current legislation, with goals that are being met, since 2018. Thus, in the last 5 years, it has recycled approximately 150,000 ts, transforming the recycled aggregate into rural road coverage, enforcing compliance with agenda 2030, preserving the ecosystem, and becoming a model city for other municipalities.

Keywords: Sustainability. Agenda 2030. Environmental Responsibility. Municipal Plan for Integrated Solid Waste Management.

5.3 INTRODUCTION

In 2010, Brazil generated around 33,000,000 tons of CDW per year, ABRELPE (2020), and in 2021 it generated approximately 46,000,000 ABRELPE (2021). The progressive increase in the generation of construction and demolition solid waste (CDSW) produced in the country has prompted a series of actions on the part of the federal government, including the development of CONAMA Resolution 307 of

July 5, 2002, which established guidelines, standards and procedures for the management of construction waste in municipalities (BRASIL, 2002).

This study was carried out in the municipality of Betim - MG, which belongs to the Metropolitan Region, has an area of 345.99 Km^2 and a population of approximately 450,000 people (IBGE, 2021).

In 2015, the municipality's Department of the Environment drew up the Municipal Plan for Integrated Solid Waste Management (PMGIRS), following the guidelines of Conama Resolution 307 of July 5, 2002, and based on the PMGIRS, actions began to be implemented to reduce the impacts on the environment caused by the waste generated by the construction and demolition sector (PMGIRS, 2015).

According to the PMGIRS (2015) and data collected in 2009, the management of CDW in the municipality, like the majority of Brazilian cities, was not providing adequate management of Construction and Demolition Waste/ Resíduo de Construção Civil (CDW), nor was it following CONAMA resolution 307 and ABNT standards.

In 2015, the municipality, through the Secretary for the Environment, set up 14 Small Volume Collection Units (URPV). Each URPV has an average of three buckets with a volumetric capacity of $8m^3$, and it is estimated that each URPV generates around 3,000 tons of rubble per month. The waste was collected and sent to Sorting and Transshipment Area (ATT) of the Construction and Demolition Waste/Construction Waste Recycling Plant (PMGIRS, 2015).

The PMGIRS set targets for the reuse and recycling of MSW in 2015, to be achieved by 2020. In addition, one of the goals of the municipality's plan was to create municipal legislation for better management and, in 2018, LAW No. 6412, OF SEPTEMBER 12, 2018 was created, which provides for the Municipal Basic Sanitation Policy (PMB, 2018).

The management of public policies in the municipality is carried out by ECOS - Empresa de Construção, Obras, Serviços, Projetos, Transporte e Trânsito de Betim (Betim Construction, Works, Services, Projects, Transportation and Transit Company), which was transformed into a municipal authority in 2018 by law 6381/2018, CMB (2018).

This study aims to contribute to theoretical and practical discussions on the subject of Construction and Demolition Waste/Construction Waste in the scientific environment, as well as encouraging municipal managers to practice the management of CDW in accordance with CONAMA resolution 307, ABNT and other regulations to preserve the environment, by evaluating the management of CDW in the city of Betim from 2015 to 2022.

With this in mind, the aim of this article was to diagnose the management of CSR destined for the plant in Betim.

5.4 METHODOLOGY

This work used the applied research method, of a descriptive nature, from a qualitative approach, through a case study. The study scenario was the municipality of Betim - MG, located 30 km from Belo Horizonte, in the metropolitan region of Minas Gerais - one of the municipalities in the southeast region with the highest urban growth according to IBGE, (2020).

The Betim Municipal Integrated Solid Waste Management Plan was read to understand the management situation and the proposals listed. After reading it, data was collected on waste generation in 2015 from the PMGIRS and the targets described in the document for dealing with CDW.

According to Marconi and Lakatos (2017), the quantitative-descriptive method aims to delineate or analyze the characteristics of facts or phenomena, evaluate programs or isolate main or key variables. In this context, the CSR generated was collected from 2018 to 2022 - annually, monthly and daily - with managers from the company ECOS - Empresa de Construções, Obras, Serviços, Projetos, Transporte e Trânsito de Betim.

A sampling study was carried out on the amount of Construction Waste generated in the municipality of Betim between 2015 and 2022 and the generation of CDW and recycling of this waste between 2018 and 2022. After this survey, graphs were drawn up to better understand the data. The data was processed using descriptive statistics, namely means and percentages, in order to compare the waste generated between 2015 and 2022.

5.5 RESULTS AND DISCUSSION

According to ABRELPE's Panorama (2021), in 2021 there was a 5.5% increase in CSW generated in Brazil compared to the previous year. Across the country, around 47 million tons were collected per year, with the Southeast accounting for 34% of this total, generating around 24 million tons of waste per year.

The municipality of Betim has a significant amount of solid waste, and according to the Betim Municipal Integrated Solid Waste Management Plan (PMGIRS), a total of 82,152,120 kg/year was generated in 2014, and it is estimated that 23% of the total waste generated in the municipality between 1997 and 2014 is DSW, PMGIRS (2015).

In order to get the plant up and running, the Betim City Council (PMB) had to buy new equipment and change the machinery, since, despite having been set up in 2011, the plant had never worked until 2018. According to the municipality, it had problems such as undersized equipment and errors in its assembly, ECOS (2018).

According to Betim's Municipal Plan for Integrated Solid Waste Management (PMGIRS), in 2015, a period when recycling was not yet practiced in the municipality, around 103 tons of CDW were generated per day. By 2022, the plant would be receiving an average of 150 tons per day for recycling.

Graph 1 - The plant's daily production control for the months of January to June 2022.

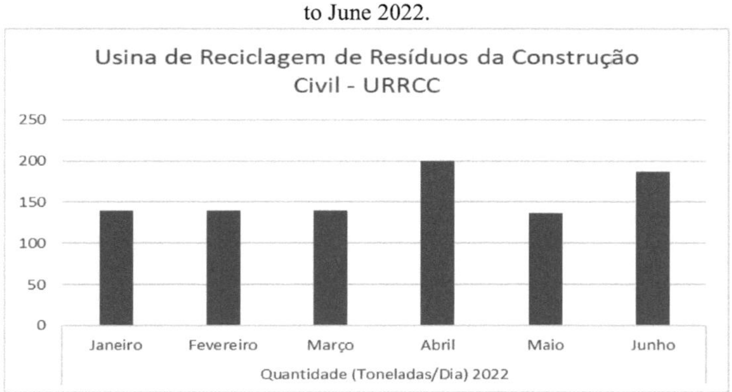

Source: PMB adapted by Diniz R.G.N, 2022

The average calculation took into account 20 days of monthly production due to downtime for preventive and corrective maintenance, as well as seasonal events.

In the survey from January to July 2022, the average demand was 150 tons/day considering 20 days/month of work at the plant, totaling approximately 3,000 tons/month. With this monthly average over 12 months, we will have 36,000 tons generated in the city of Betim - MG.

Graph 2 - The plant's monthly production control for the months of January to June 2022

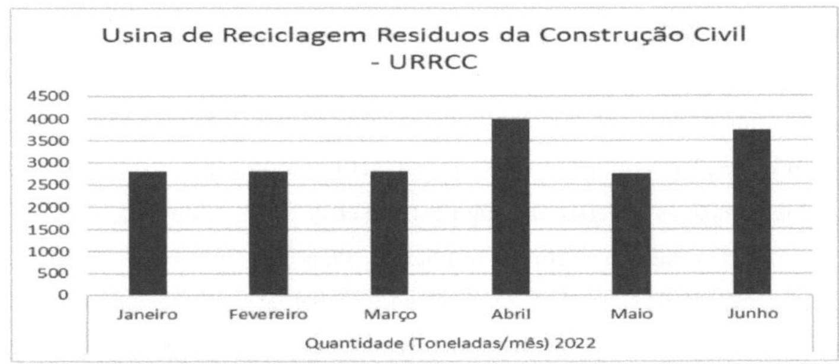

Source: PMB adapted by Diniz R.G.N, 2022

Betim City Hall, through the company ECOS, has been crushing the MSW generated in the municipality since 2018. It began with a capacity to recycle 2,500 tons and by 2021, it was crushing an average of 2,800 tons per month.

The plant has maximum crushing capacity with the potential to increase production as well as the collection of CDW.

Graph 3 - Annual production control of the plant, from its inauguration until 2022.

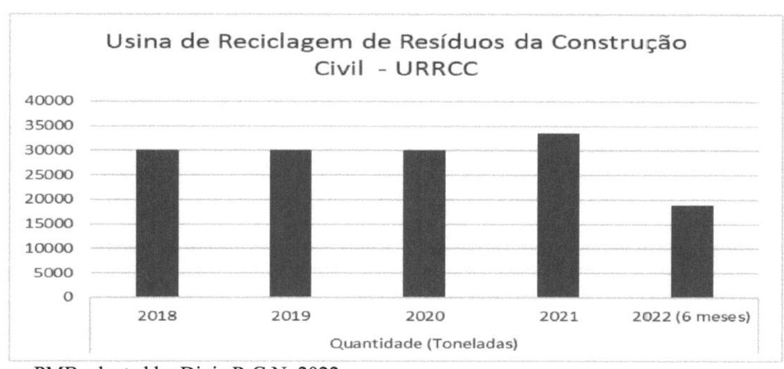

Usina de Reciclagem de Resíduos da Construção Civil - URRCC
40000
35000
30000
25000
20000
15000
10000
5000
0
2018
2019
2020
2021
2022 (6 meses)
Quantidade (Toneladas)
Source: PMB adapted by Diniz R.G.N, 2022

According to the DSA/ECOS Usina de Reciclagem de Resíduos da Construção Civil - URRCC in the city of Betim - MG in 2018, it recycled 30,000 tons per year at its maximum capacity and in 2022 it reached 33,000 tons per year, with an increase forecast according to the demand applied.

The daily DCSRs generated between 2015 and 2022 represent a percentage increase of 45.63 percent in just seven years, with an average annual increase of 6.5 percent, as shown in figure 15.

Graph 4 - CSR generated between 2015 and 2022 in the city of Betim- MG

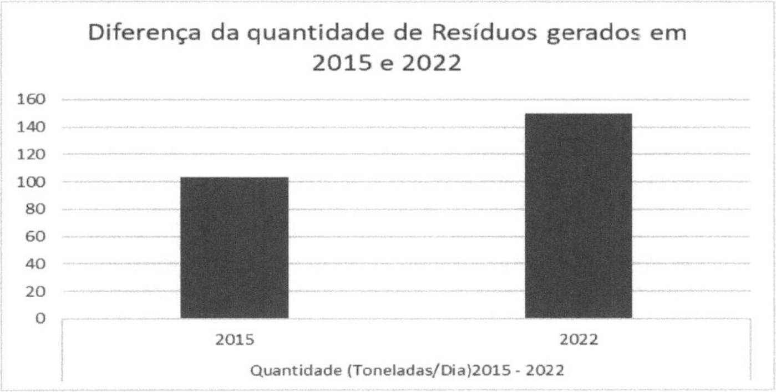

Diferença da quantidade de Resíduos gerados em 2015 e 2022
160
140
120
100
80
60
40
20
0
2015
2022
Quantidade (Toneladas/Dia)2015 - 2022
Source: PMB adapted by Diniz R.G.N, 2022

If we forecast an average percentage from 2022 to 2030, with the construction industry maintaining the scale of production of the last 7 years, there will be a 52% increase in CDW in the city of Betim, which indicates a need to keep the plant running to prevent this waste from being disposed of incorrectly, causing damage to the environment.

The city of Betim stands out in its solid waste management by meeting the goals of the PGRS. Since 2015, the city has removed an average of 250,000 tons of solid waste from possible clandestine landfills and dumps over a period of 7 years. By 2030, we can estimate that this figure will double, meaning that at least 520,000 tons will be disposed of correctly without degrading the environment, ECOS (2018).

It should be emphasized that construction is one of the sectors that most degrades the environment, impacting biodiversity, polluting the atmosphere, reducing water reserves, and especially generating waste, especially when the work is not done with proper planning Laruccia (2014).

To avoid environmental degradation due to the incorrect disposal of CDW, the generation of this waste, meeting consumption needs without compromising the environment and future generations, complying with legislation and fulfilling the proposals of Brazil's 2030 agenda. According to Gnecco (2018), new technologies have been advancing to minimize these environmental impacts, such as BIM (*Building Information Modeling) technology.* However, the tool is still positioned as a technology with a high price tag: BIM software costs an average of R$1,400.00/month for each license, in addition to the need to hire a consultant to implement and use it. Therefore, small companies and property owners - who are going to demolish, build or renovate - still do not have access to technologies such as BIM and this audience represents a large portion of waste generators, Autodesk (2022).

The tool is a technological platform applied to civil construction that makes it possible to model the projects and details of the specifications of the installations or the building, avoiding waste and costs in construction. CATELANI (2016).

Marinho (2017) proposes that BIM is capable of managing and storing information and data throughout the entire life cycle of the project, from the conception and conceptualization of the idea, project development, construction, and also after the project is finished, in its use phase.

Brito and Ferreira (2015) define BIM as a design methodology that supports the integration of information throughout the stages of the design process, project management and construction. Therefore, more than the practice of recycling, it is necessary to encourage the reduction of CSR generation.

That said, the purpose of this study is to demonstrate that compliance with Brazil's 2030 agenda is possible through compliance with legislation, as well as to encourage other municipalities to carry out the correct management of solid construction waste, reusing it once it has been recycled. Since the materials resulting from the recycling process can be used in the manufacture of bricks, blocks, half-wires and others, thus avoiding environmental degradation, leaving future generations with a preserved ecosystem guaranteeing them a quality of life, as well as proposing the reduction of CDW through research into new technologies.

5.6 FINAL CONSIDERATIONS

The research carried out a diagnosis of the management of CDW in the city of Betim with annual, monthly and daily data since the preparation of the PMGIRS, which took place in 2015, also brought concepts about Construction and Demolition Waste/Construction Waste in accordance with Conama Resolution 307 of July 5, 2002 in the scientific environment, in addition to encouraging municipal managers to practice the management of CDW in accordance with CONAMA Resolution 307, ABNT and other regulations to preserve the environment, evaluating the management of CDW in the city of Betim from 2015 to 2022.

The municipality of Betim follows the rules to meet its targets for recycling solid construction and demolition waste, and the city has worked to reduce environmental impacts.

With the environmental objective in mind, the amount of hazardous waste that would otherwise have been disposed of in "dumps" or clandestine landfills has been considerably reduced with the creation of the plant.

In order to avoid these "Bota fora", clandestine disposals that degrade the environment, it is necessary to carry out special studies on the management of CSR as well as promoting recycling, meeting consumption needs without harming the environment, complying with legislation and enforcing the 2030 Agenda in Brazil.

Recycling MSW is an important step towards achieving a sustainable ecosystem, and in partnership with public and private authorities, dealing with waste becomes more accessible. In the case presented, it was extremely important to create an autarchy in the municipality to reduce bureaucracy in public management.

Complying with CONAMA/307 by drawing up a municipal integrated solid waste management plan is the first step towards effective solid waste management and the way to ensure compliance with the 2030 agenda. As the research showed, the initiative to draw up the PMGIRS was what led the municipality to achieve the targets set.

Research is needed into new technologies to prevent the generation of solid construction and demolition waste.

Encouraging other municipalities to recycle through examples such as the city of Betim - MG, is important to reduce bureaucracy and demonstrate that it is possible for Brazil to reach 2030 as a country that has managed to fulfill its agenda.

Ultimately, encouraging the adoption of recycling techniques and good CSR management practices to preserve the environment is a way of guaranteeing a better life for current and future generations.

5.7 REFERENCES

ABRELPE - Brazilian Association of Public Cleaning and Special Waste Companies In: **Panorama 2021**. [S. l.], 2021. Available at: https://abrelpe.org.br/panorama/. Accessed on: Jan. 8, 2022.

AUTODESK (USA). **Revit: BIM software for designers, builders and developers. In: Revit:** BIM software for designers, builders and developers. 1.0-RC2. [S. l.], 2022. Available at: https://www.autodesk.com.br/products/revit/overview?term=3YEAR&tab=subscription. Accessed on: July 25, 2022.

BRAZIL - CHAMBER OF DEPUTIES. Ministry of Foreign Affairs, Environment Division, with the collaboration of the United Nations Environment Program, UNEP. Parliamentary Action, 56, 1995. 56. **United Nations Conference on Environment and Development**, Brasília-DF: Documentation and Information Center, Publications Coordination, v. 1, p. 472, 1995.

BRAZIL - National Secretariat for Environmental Sanitation. Network for training and technological extension in environmental sanitation. ReCesa, 2007. Resíduos sólidos: PGRSU -Plano **de gestão de resíduos Sólidos urbanos:** : guia do profissional em treinamento, Belo Horizonte- MG: Artes Gráficas Formato LTDA, v. 1, p. 100, 2007.

BRITO, D. M. de; FERREIRA, E. de A. M. **Evaluation of strategies for representing and analyzing construction planning and control using 4D BIM models**. Ambiente Construído, Porto Alegre, v. 15, n. 4, p. 203-223, 2015.

BETIM CITY COUNCIL. Ordinary Law No. 6381, of June 29, 2018. Provides for the transformation of the Betim construction, works, services, projects, transportation and traffic company - ecos into a municipal public authority, and makes other provisions. **Ordinary Law No. 6.381/2018**, Betim, 2018

CATELANI, Wilton Silva (Brazil). Cbic - **Brazilian Chamber of the Construction Industry.**

Coletânea Implementação do BIM Para Construtoras e Incorporadoras: Fundamentos BIM - Available at: https://cbic.org.br/inovacao/2017/10/18/coletanea-bim/ Accessed on: July 12, 2022.

ECOs - Empresa de Construções, Obras, Serviços, Projetos, Transporte e Trânsito **Annual report, 2019**. Betim City Hall (Minas Gerais). Betim -MG, 2019.

- Empresa de Construções, Obras, Serviços, Projetos, Transporte e Trânsito **Annual report, 2020**. Betim City Hall (Minas Gerais). Betim -MG, 2020.

- Empresa de Construções, Obras, Serviços, Projetos, Transporte e Trânsito **Annual report, 2022**. Betim City Hall (Minas Gerais). Betim -MG, 2022.

FRASSON, S.. Rubble recycling plants as agents in the valuation of waste generated by civil construction, [s. L.], 12017.

GNECCO, verônica. **bim for waste management in civil construction: case study at the cras construction site in the municipality of biguaçu/sc**. 2019. 150 f. TCC (Bachelor's Degree) - Technology Center of the Federal University of Santa Catarina, Santa Catarina, 2018.

IBGE - Brazilian Institute of Geography and Statistics. Population 1980. In: Federal Government. IBGE. Population. [S. l.], 2021. **Total population - 1980 - 2010**. Available at: https://brasilemsintese.ibge.gov.br/populacao/populacao-total-1980-2010.html. Accessed on: May 1, 2022.

LARUCCIA, Mauro. **Sustainability and Environmental Impacts of Construction**. Article, [s. l.], 2014.

MARCONI, Marina d; LAKATOS, Eva. **Fundamentals of scientific methodology**. 8. ed. actual. São Paulo: Atlas, 2017. ISBN 978-85-970-1076-3.

MY Municipality. In: **Data from Betim**. 1.0-RC2. [S. l.], 2015. Available at: https://meumunicipio.org.br/perfil-municipio/3106705-Betim-MG?exercicio=2015. Accessed on: June 14, 2022.

PMGIRSB - **Betim municipal integrated solid waste management plan.** PMB - Betim City Hall (Minas Gerais). 2015, Betim -MG, p. 1-128, 2015.

BETIM CITY HALL. **Law No. 6412,** of September 12, 2018. Provides for the municipal basic sanitation policy and makes other provisions. Betim Minas Gerais, September 12, 2018.

SEMMAD - Municipal Secretary for the Environment and Sustainable Development - **Municipal Solid Waste Management Plan.** PMB - Betim City Hall (Minas Gerais). 2015, Betim -MG, p. 1-128, 2015.

6 CHAPTER III

CASE STUDY: STUDY OF THE CURRENT SCENARIO OF THE SOLID CONSTRUCTION WASTE RECYCLING PLANT IN THE MUNICIPALITY OF BETIM/MG

6.1 SUMMARY

Every enterprise, whether public or private, depends on good management and economic viability in order to stay in business. The study from a financial, environmental and social perspective plays a crucial role in decision-making, making projects successful and sustainable. That said, the aim of this article was to analyze the management of the Civil Construction Solid Waste Recycling Plant (RSCD) in Betim, Minas Gerais, and to identify its reality from a financial, environmental and social point of view. The research used the applied, descriptive and qualitative research method, by means of a case study. The data was obtained through interviews and questionnaires with the waste management board of the Betim Construction, Works, Services, Projects, Transportation and Transit Company (ECOS) and on-site visits to the plant. We analyzed how the plant has been managed from its opening to the present day, using the data provided by the management company. As a result, the plant did not prove to be financially attractive in the current management molds, operating with a monthly negative balance of R$ 1,948,486.89, however, when evaluating the environmental and social issue, it explains its importance to society and the environment.

Keywords: Recycled aggregate. Recycling. SDGS. Agenda 2030.

6.2 ABSTRACT

Every enterprise, whether public or private, depends on good management and being economically viable to remain in the market. Studying from a financial, environmental and social perspective plays a crucial role in decision-making, making projects successful and sustainable. That said, this article aimed to analyze the management of the Civil Construction Solid Waste Recycling Plant in Betim-MG, as well as identify its reality from a financial, environmental and social point of view. For the research, the applied, descriptive and qualitative research method was used, through a case study. The data were obtained through interviews and questionnaires with the waste management department of the company (ECOS) and through on-site visits to the plant. We analyzed how the Plant is managed from its opening to the present day, using data provided by the management company. As a result, the plant did not prove to be financially attractive under the current management model, operating with a negative monthly balance of R$ 1,948,486.89, however, when evaluating the environmental and social issue, it explains its importance to society and the environment.

Keywords: Recycled aggregate. Recycling. SDGs. 2030 Agend.

6.3 INTRODUCTION

Human actions and the infrequent use of natural resources in recent decades have contributed to the imbalance and devastation of the environment, causing irreparable damage to the ecosystem and tragic consequences.

According to Cidreira-Neto; Rodrigues (2017), man uses nature's resources in an exploitative way, without foreseeing the problems that this attitude can cause, generating problems of a social and/or environmental nature, which can have their effects quickly or over a long period of time.

The construction sector, despite its importance in economic development, is one of the sectors that most degrades the environment throughout its production chain. According to Pinto (2013), the biggest related problem is the inadequate disposal of waste, compromising the lifespan of the environment. According to Pascoalina Filho and Graudenz (2012), improperly disposed CSW can pollute the soil, degrade urban landscapes and pose a severe threat to public health.

The United Nations Organization - UN , through a series of conferences and meetings, starting in 1972, began to give visibility to Sustainability, and from these conferences emerged the 2030 agenda that guides government actions to ensure the effective development of international environmental policies. Brazil is one of the countries facing the challenge of meeting the 17 Sustainable Development Goals - SDGs by 2030.

One way in which Brazil is taking part in the 2030 agenda is through legislation. In 2002, CONAMA Resolution 307/2002 was drawn up to establish guidelines for the management of CDW (BRAZIL, 2002).

With the visibility of sustainability, the Concept of the *Triple Bott Line (TBL*), also known as the sustainability tripod, gained repercussions at the end of the 1990s, integrating the economic, social and environmental aspects. According to Costa and Ferezin (2021), the success of today's organizations is linked to the TBL, so the management of CSR must meet these criteria, and according to the authors, economic sustainability is considered the first pillar of the TBL, with profit and capital as the fundamental premise.

Betim, a city in the metropolitan region of Belo Horizonte, where this study was carried out, has an area of approximately 345.99 square kilometers and a population of approximately 450,000 people, according to IBGE (2021). In 2018, it set up a recycling plant to recycle waste.

Solid Construction and Demolition Waste (URSCD) to crush the solid construction and demolition waste generated in the municipality.

The URSCD was acquired in 2011 through a partnership between the municipality, Bancos do Brasil and Banco Nacional Desenvolvimento Econômico e Social -BNDES , but it only started operating in 2018, when new equipment was acquired to support the crushing capacity of the MSW generated in the city, according to ECOS (2022). ECOS was declared a municipality in 2018 and is the manager of Betim's public policies and is currently directly responsible for the plant, which began with a capacity to recycle 2,500 tons and in 2021 was crushing an average of 2,800 tons per month (ECOS, 2022).

The municipality's intention in setting up the plant was to offer sustainable alternatives for the treatment of MSW.

This research sought to carry out an analysis of the plant's management, as well as analyzing its financial, environmental and social aspects.

6.4 METHODOLOGY

The work was carried out using an applied, descriptive, qualitative research technique, through a case study.

The study site was the Solid Construction Waste Recycling Plant located in the municipality of Betim - MG, the city of Betim is located 30 km from Belo Horizonte, in the metropolitan region of Minas Gerais - one of the municipalities in the southeastern region of Brazil.

Figure 2 - Map of Betim, metropolitan region of Belo Horizonte - MG

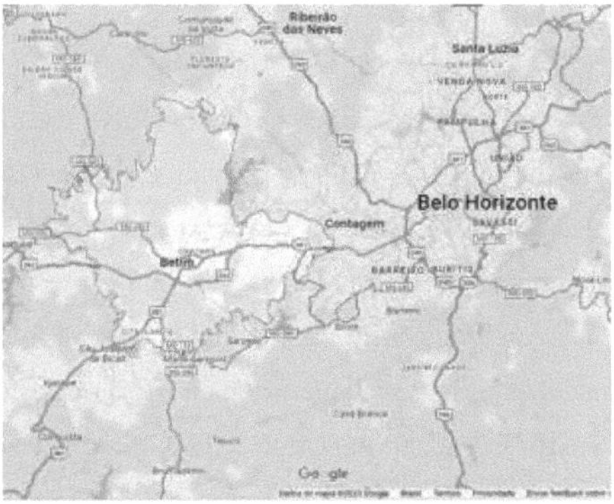

Source: *Google Maps* (2023).

The main stages of this research were:

a) Bibliographical research;
b) Visits to the recycling plant in Betim;
c) Questionnaire sent to the actors involved in the process;
d) Analysis of the Plant's management; and
e) Study of the financial, environmental and social reality of the project.

For the literature review, research was carried out on academic papers on the subject of URSCD and RSCD management on the CAPES journals and academic *google sites* with the keywords: Construction and demolition waste management; Feasibility of construction waste recycling plants, recycling plants for solid construction waste, research was also carried out on *websites* and representative bodies, including SINIS and Gov.br.

In order to analyze the plant's management, visits were made to the fields on working days during the plant's opening hours. The visits were accompanied by the direct manager representing the town hall, the plant supervisor and the supervisor. At the same time, we were shown the plant's monthly measurements. All the data relating to the URSCD in this research was provided by the management of the ECOS company through the aforementioned channels.

The analysis of the financial, environmental and social management of the URSCD in Betim was carried out by analyzing the monthly measurements sent by the company that manages the plant. The unit of measurement was calculated from tons to $1m^3$, considering that for every 1.45 tons of recycled aggregate, we had $1m^3$. After the conversion, we researched the price of aggregate and chose the official data from National System for Researching Civil Construction Costs and Indices (SINAPI).

The financial analysis calculations took into account the plant's production capacity in M3, the monthly cost and the value of $1m^3$ of recycled aggregate.

6.5 RESULTS AND DISCUSSION

According to Federal Law 12.305/2010, the following order of priority must be observed in waste management: non-generation, reduction, reuse, recycling, treatment and environmentally correct final disposal (BRASIL, 2010).

Recycling plants help to reduce the environmental impact of construction waste. In this way, the plant not only serves as a final destination for the waste, but also as a source of inserting the recycled aggregate back into the production chain.

According to ABRECON - The Brazilian Association for the Recycling of Construction and Demolition Waste - highlights the alternative applications for recycled construction waste, as shown in table 11 below:

Table 11 - Alternatives for using recycled aggregates

PRODUTO	USO RECOMENDADO
Areia reciclada	Argamassas de assentamento de alvenaria de vedação, contrapisos, solo-cimento, blocos e tijolos de vedação.
Pedrisco reciclado	Fabricação de artefatos de concreto, blocos de vedação, pisos intertravados, manilhas de esgoto, dentre outros.
Brita reciclada	Fabricação de concretos não estruturais e obras de drenagens.
Bica Corrida	Obras de base e sub-base de pavimentos, reforço e subleito de pavimentos, regularização de vias não pavimentadas, aterros e acerto topográfico.
Rachão	Obras de pavimentação, drenagem e terraplenagem.

Source: ABRELPE, 2018

The URSCD under study is located in Betim, Minas Gerais, and has a total area of 20,000 m² and an operating area of 1,800 m2 , distributing equipment and handling areas. According to ECOS (2022), there is no plan or sketch of the URSCD.

When acquiring the land, a number of points were taken into account, such as access routes, noise and vegetation. The local hydrology was not taken into account, as the land is free of water bodies.

The URSCD was acquired through a partnership between the Municipality of Betim and Banco do Brasil and BNDS in 2011, but never operated until 2018, when an investment of R$ 790.00 was made with the aforementioned banks to purchase and adapt new equipment to adjust its operation to the demands of the city.

The plant processes the waste and the product from this processing is classified as gravel 1 and 2 depending on its particle size and is stored in the open by type of recycled aggregate.

In the figure below, photos of the plant as described:

A. Crushing area;

B. Crushing conveyor;

C. Granulated aggregate;

D. Gravel sample 2 - recycled aggregate from the plant;

E. Aggregate bucket; and

F. Production area.

Figure 3 - Recycling plant for solid construction waste in Betim - MG

Source: Diniz R.G.N, 2023.

The stage of the waste processing process begins with the collection of the debris, which is stored together regardless of its origin, and ends with the recycled aggregate being returned to the production chain; the recycled aggregate is used as road cover on rural roads.

Figure 3 shows the stages of the plant's production process, using the VSM model. To draw up the flow map below, a pull system was used, as the plant has a maximum production capacity of 2800t/month. According to the *Lean Institute Brazil* (2024), pull production tries to eliminate excess production.

Figure 4 - Usida process using VSM (*Value Stream Mapping*) model

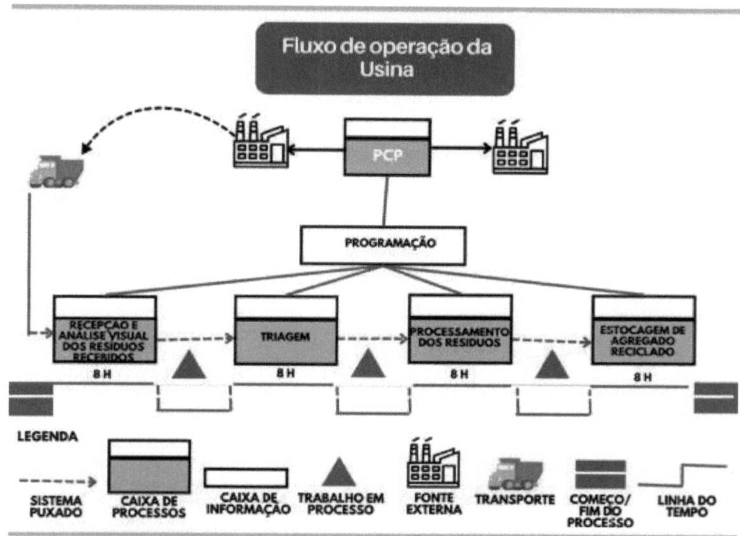

Source: Prepared by Diniz R.G.N, 2023.

VSM (*Value Stream Mapping)* is a *Lean manufacturing* tool that helps us to see the production chain, to improve the process and to diagnose it so that, if necessary, we can create action plans that reduce unproductive time, unnecessary costs, increase profits and production capacity, among other advantages of the tool. According to ROTHER; SHOOK (2003); SHOUet *al.* (2017) The VSM helps to understand what adds value or not, from the supplier to the end consumer.

Currently, URSCD is managed by VIASOLO Environmental Engineering, which receives R$ 36,520,597.92 per year for the management of operations through a bidding contract, which is expected to end in 4 years after the start of activities.

According to ECOS (2023), the URSCD has a monthly budget plan of approximately R$3,000,000.00/month. The administrator of the Betim URSCD operation, Viasolo Engenharia Ambientais, has had a contract with the Betim City Hall (PMB) since 2019.

Payment is made by monthly measurement, and the annual value of the contract is R$36,520,597.92, while the monthly value planned in the contract is R$3,043,383.16.

The plant's production capacity is 28,000 tons (this is the monthly production agreed in the contract), and every month the URSCD has to deliver 28,000 tons to the municipality for disposal in rural road cover. The companies (e.g. MRV) send the rubble (construction waste to the plant), free of charge. The rubble arrives at the plant and is sorted to check for contaminants. If there are no contaminants, the rubble goes to the crusher (a machine that processes the rubble for recycling). The rubble is transformed into recycled aggregate considered type 1 and 2 gravel.

The price of recycled aggregate gravel type 1 and 2 is R$56.70 per cubic meter, according to an analytical table from the National System for Researching Construction Costs and Indices (SINAPI).

A study carried out by cooperativa de pesquisas tecnológicas industriais (CPTI) - showed the average conversion rate considering that for every 1.45 tons of gravel 1,2,3 we have 1m3, with a tolerance range of approximately 5%, so the financial analysis was based on the values of recycled aggregates practiced on the market, initially converting the units of measurement.

1,45 t. ------------------- 1m^3

28.000 t. --------------------X

Table 1 - Annual/monthly plant revenue

Monthly/annual revenue				
Monthly production (m^3)	Added value R$	Monthly revenue R$	Months	Annual revenue R$
19.310,34	56,7	1.094.896,27	12	13.138.755,33

Source: Prepared by Diniz R. G. N., 2023.

Considering that PMB makes an annual investment of R$36,520,597.92 and generates annual revenue of R$13,138,755.33, if the plant does not generate any revenue other than the sale of gravel to the municipality, it will operate with a monthly negative balance of R$1,948,486.89, making it economically unviable under these conditions. It should be noted that for this study, the initial investment and losses during the process were disregarded.

The process of recycling CSR in Brazil is developing and, in order to make it a reality, it is crucial, according to Toledo, 2013, that the practice is based on the three fundamentals of sustainability: being environmentally correct, socially fair and economically viable.

As it is a plant, it receives rubble and disposes of it in an environmentally appropriate manner, at no cost to construction companies, which dispose of their waste on site, thus avoiding incorrect disposal in areas such as dumps or garbage dumps. Its operation contributes to the conservation of natural resources, which are replaced by recycled aggregates. In addition, through the practice of recycling, the plant carries out environmental education, demonstrating the importance of maintaining a healthy environment. The plant therefore plays a fundamental role in good environmental practices.

The plant directly employs eight people, as well as indirect collaborators such as city hall managers, dumpers, drivers, among others. In addition, the disposal of recycled aggregate on rural public roads for structural improvements and its economic stimulus has a significant impact on society.

6.6 FINAL CONSIDERATIONS

When analyzing the plant's management, it became clear that the project does not have a prior environmental license or an installation license, only an operating license. There has been no popular acceptance survey, no opening statement, the plant has no management plan, there is no planning or estimate of costs and acquisitions, and there is no plan for intervening in environmental impacts.

Specific costs, such as water, electricity, personnel and administration, were not made available to the research because they are confidential, according to the plant manager, and so it was not possible to carry out the Payback for the project.

The survey showed that financially it has a considerable deficit in its operation.

In terms of the environmental and social aspects, the research showed that the plant plays a fundamental role in contributing to both.

In order to improve the financial situation, it is necessary to reduce costs, draw up a plan and implement lean management practices to add value to the material processed. Make offers and encourage the purchase of recycled aggregates from neighboring municipalities, increasing production capacity. Enforcing the disposal of rubble by creating municipal laws to prevent dumping and making it compulsory to dispose of it at the URSCD.

Encouraging companies to recycle on site for reuse, ensuring that this waste is not disposed of in inappropriate places and providing tax incentives to receive granulated aggregate in return. The plant proved to be economically unviable, which does not apply to technical and environmental viability, where further studies are needed to prove this.

6.7 APPENDIX

Some of the questions in the questionnaire were:

1. There is an opening term
2. There is planning and cost estimation
3. There is a budget determination
4. How large is the plant?
5. Total land area
6. There is a Plan for the Treatment of Environmental Impacts caused by the operation of the Plant
7. Has an environmental license

6.8 REFERENCES

ABNT - BRAZILIAN ASSOCIATION OF TECHNICAL STANDARDS. NBR 10.004 - **Solid waste** - Classification. 2004.

ABNT - BRAZILIAN ASSOCIATION OF TECHNICAL STANDARDS. NBR 15112- **Construction waste and bulky waste** - Transshipment and sorting area - Guidelines for design, implementation and operation. 2004.

ABNT - BRAZILIAN ASSOCIATION OF TECHNICAL STANDARDS. NBR 15113- **Solid construction waste and inert waste** - Landfills - Guidelines for design, implementation and operation. 2004.

ABNT - BRAZILIAN ASSOCIATION OF TECHNICAL STANDARDS. NBR 15114 - **Solid waste**

ABNT - BRAZILIAN ASSOCIATION OF TECHNICAL STANDARDS. NBR 15115 - **Recycled aggregates from solid construction waste** - Paving layers - Procedures. 2004.

ABNT - BRAZILIAN ASSOCIATION OF TECHNICAL STANDARDS. NBR 15116 - **Recycled aggregates from solid construction waste** - Use in paving and concrete preparation without structural function - Requirements. 2004

BRAZIL. **CONAMA Resolution No. 348 of August 16,** 2004. Amends CONAMA Resolution 307, of July 5, 2002, including asbestos in the class of hazardous waste. Brasília, 2004.

BRAZIL. Civil House. Law No. 12.305, of August 2, 2010. **Establishes the National Solid Waste Policy; amends Law No. 9.605, of February 12, 1998**; and makes other provisions. Brasília: DOU, 2010.

CIDREIRA NETO, Ivo Raposo Gonçalves; RODRIGUES, Gilberto Gonçalves. **The man-nature relationship and the limits to sustainable development.** Revista Movimentos Sociais e Dinâmicas Espaciais, Recife, v. 6, n. 2, 2017.

COSTA, E., Ferezin, N. B. **ESG (Environmental, Social and Corporate Governance) and communication: the sustainability tripod applied to globalized organizations.** Revista ALTERJOR, São Paulo, v. 2, n. 24, p. 79-95, jul./dez, 2021

PASCHOALIN FILHO, J. A.; Graudenz, G. S. **Irregular Disposal of Construction and Demolition Waste (CDW) and its Impact on Public Health**. Environmental & Social Management Journal, v. 6, n. 1, 2012.

PINTO TP. **Methodology for the differentiated management of solid urban construction waste**. Thesis (Doctorate) - Polytechnic School of the University of São Paulo. Department of Civil Construction Engineering, São Paulo, 2013, 189p.

SHOU, W.; WANG, J.; WU, P.; WANG, X.; CHONG, H-Y. **A cross-sector review on the use of Value Stream Mapping. International Journal of Production Research**, v. 55, n. 13, p. 3906-3928, 2017

TOLEDO, M. C. C. de. **Recycling in Civil Construction**. 2013. Available at: <http://www.guiadecorar.com.br/posts/visualiza/639>. Accessed on: June 25, 2015

7 CHAPTER IV

TECHNICAL-SCIENTIFIC REPORT ON THE THEORETICAL POTENTIAL FOR GENERATING RECYCLABLES IN THE SELECTIVE COLLECTION OF SOLID CONSTRUCTION AND DEMOLITION WASTE IN THE MUNICIPALITY OF BETIM/MG

7.1 SUMMARY

Urban waste management is a challenge for Brazilian cities, especially medium-sized and large ones, mainly due to the scarcity of human and financial resources. The implementation of a selective collection service for construction and demolition waste (CDW) has the potential to improve waste management. This chapter will study the theoretical potential for generating recyclables (PTGR), which is the starting point for planning the municipal recycling policy, becomes the indicator of the potential for recyclable waste that is generated by the population in the municipality and sets operational targets for the CSW recycling plant. Thus, through a literature review and the collection of information from public databases, we intend to analyze the PTGR/RSCD of the city of Betim. The methodology PDCA (*Plan, Do, Check, Action*) will be used for diagnosis and continuous corrections of any problems that may arise and associated with the theoretical potential for generating recyclables (PTGR), making a future projection of 10 years. This process will be implemented gradually, so that RSCD collection will begin in a specific neighborhood in the city. And finally, proper disposal to ensure that the collected MSW is sent to the MSW recycling plant.

Keywords: Selective collection. Construction and demolition waste. Theoretical potential for generating recyclables.

7.2 ABSTRACT

Urban waste management is a challenge for Brazilian cities especially for medium and large sized companies due mainly to the scarcity of human and financial resources. The implementation of the selective collection service for Solid Waste from

Construction and Demolition (CDW) has the potential to improve waste management. Generating income and valuing the recycled raw material in this chapter will be studied the theoretical potential for the generation of recyclables (PTGR), which is the starting point for the planning of the municipal recycling policy, becomes the indicator of the potential of recyclable waste that is generated by the population in the municipality and defines operational goals of the CDW recycling plant Thus, through the literature review and collection of information in a public database, it is intended to analyze the PTGR/CDW of the city of Betim. The PDCA methodology (Plan, Of, Check, Action) for continuous diagnosis and correction of any problem that may arise and associated with the theoretical potential for the generation of recyclables (PTGR), carrying out a future projection of 10 years. This process will have a gradual rollout so that RSCD collection will be started in a specific neighborhood of the city. and finally, the proper disposal to ensure that the CDW collected is sent to the recycling plant for this waste.
Keywords: Selective collection. Construction and demolition waste. Theoretical potential for the generation of recyclables.

7.3 INTRODUCTION

Although it is one of the objectives of Law 12.305, it is difficult to carry out this kind of management in large cities, as there are dump sites and a lack of public policies associated with financial and human resources (PEREIRA; SOUZA, 2023; BRASIL, 2010). As a result of this context, the importance of integrated waste management is clear.

Recycling MSW is fundamental in changing the globally applied linear model to a circular economy, since it connects the ends of the chain and thus makes reverse waste logistics operational. The results presented by calculating the theoretical potential for generating recyclables reflect the environmental importance of investing in the recycling of CDW, since it reduces the extraction of natural resources through the recovery of these materials and also landfills through detour via selective collection to the UVR s (PGRS/PTI, 2022).

In Brazil, at least 75.1% of municipalities have some sort of selective collection initiative, but in most cases these are one-off initiatives that do not cover the entire population (ABRELPE, 2022).

Selective collection is one of the tools that contributes to good materials management, as it aims to separate waste at the source of generation. However, it should be noted that selective collection must be implemented in conjunction with a materials recycling program (CEMPRE, 2018).

Solid waste is still a major problem for the whole country. Many Brazilian cities still don't have an effective solid waste management system. This reality is observed in around 2,318 Brazilian municipalities, where waste is disposed of inappropriately, which corresponds to 46.2% of the cities (SNIS, 2022).

According to data from the Panorama of Solid Waste in Brazil, published in 2022 by ABRELPE, 81.8 million tons of solid urban waste were generated during the same year, meaning that each Brazilian citizen produced an average of 1,043 kilos of waste per day. Although 93.04% of all the waste generated in Brazil is collected, only 61% of this amount is properly disposed of, which is evidence of poor management (SNIS, 2022).

An example of this situation is the city of Betim-MG, located near Belo Horizonte, with a population of 41,846 (IBGE, 2022).

7.4 THEORETICAL BASIS

Implementing this type of collection generates benefits such as: reducing the use of virgin raw materials; valuing recycled raw materials; saving energy; reducing the volume of waste going to landfills; creating new businesses and generating jobs. From this perspective, it can be inferred that recyclable materials represent an available and fast-growing commodity (MIRANDA; MATTOS, 2018; CEMPRE. 2018).

Waste is classified according to its hazardousness by NBR 10.004 (ABNT, 2004) and according to its origin by Law 12.305 (BRASIL, 2010). In addition, municipalities need to manage waste and the correct destination is a problem that exists throughout the country (ABRELPE, 2022). This makes it necessary to implement measures that can reverse this situation, for example, one of the alternatives provided for in Law 12.305 (BRASIL, 2010) is Selective Collection.

In 2021, municipalities in Brazil collected more than 48,375,275 tons of CDW, which represents a 2.9% increase compared to the previous period (Figure 7).

Figure 5 - Collection of CDW by municipalities in Brazil (2021).

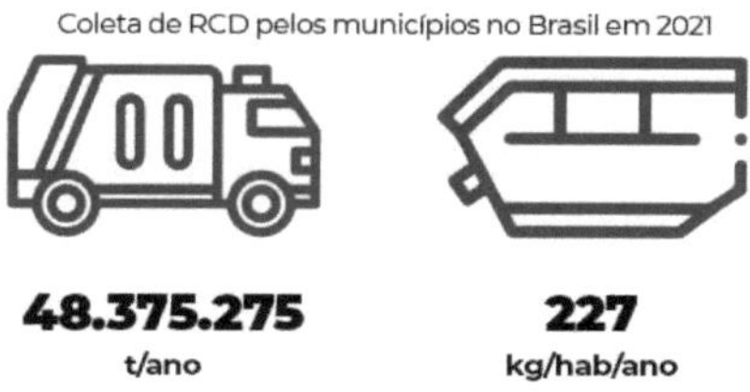

Source: ABRELBE, 2022

The amount collected per inhabitant was around 227 kg per year, equivalent to construction and demolition waste abandoned on roads and public places (Figure 20). Slightly more than half of the CDW collected in Brazil comes from the Southeast region (52%), with 279.3 kg per inhabitant per year; however, the region that stands out in terms of *per capita* collection is the Midwest, with almost 323.4 kg per inhabitant per year (Figure 8) (ABRELBE, 2022).

Figure 6 - MSW collection by municipalities in Brazil/ By region (2021)

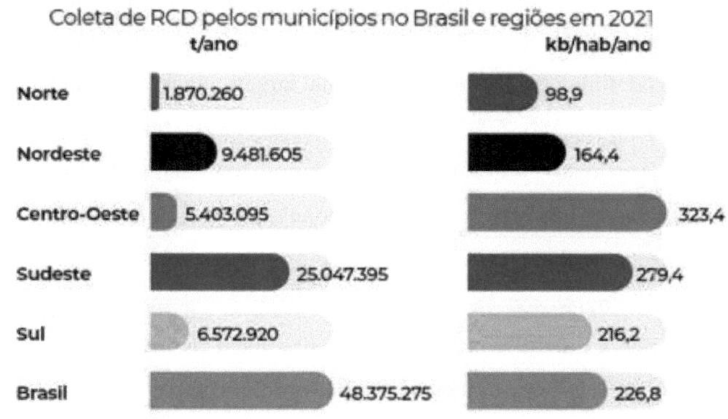

Source: ABRELBE (2022).

7.5 METHODOLOGY

With regard to the city of Betim - MG, according to information provided by the city hall, around 33,600 tons/month of waste is sent to the Construction and Demolition Waste (CDW) recycling plant, which corresponds to 11% of the city's CDW generation. Thus, on average, each citizen in the city generates 18.91 kg/inhabitant/month of waste ABRELPE (2022).

The theoretical recyclable waste potential (Eq:1), which is the starting point for planning the municipal recycling policy, becomes the indicator of the recyclable waste potential that is generated by the population in the municipality and sets operational targets for the CSW recycling plant.

PTGR = (PP) X (GPC)X (%REC) Eq. (1)

PTGR= Theoretical Potential for Generating Recyclables

PP= population;

GPC = Annual per capita generation

% REC = Percentage of CSRD recycled at the Betim - MG plant

According to the IBGE (2022), Betim has a population of 411,846 inhabitants, thus defining the project's operational targets and monitoring the recovery rate of the RSCD.

The calculation will run for 10 years until 2033, with an initial average of 18.91 kg/inhabitant/month and a projected population increase of 0.2% per year according to the national average (IBGE, 2022).

According to the Betim City Hall (PMB) and the Secretariat for the Environment (SMA) in 2022 the RSCD recycling plant recycled 33,600 kg, which corresponds to 11% recycling of the total RSCD produced in the City of Betim - MG. The study calculated that each year there will be an increase of 0.2% (if the project is implemented following the PDCA (*Plan. Do. Check. Action*).

Initially, a study will be carried out of the current situation and forecasts of waste generation in the municipality of Betim-MG, using official data from the PMB (2022).

The aim is to get to know the local reality and, above all, the difficulties faced. After this stage, a quantitative and qualitative analysis will be carried out to refine the data.

Using the information described, a system will be designed for the collection of RSCD, taking into account the peculiarities of the municipality and covering the population as a whole. To this end, guidelines will be followed:

a) Raising public awareness: Through awareness campaigns using resources such as leaflets and social media, inform residents of the details of the service to be implemented. Inform them about voluntary collection points where they can leave their RSCD. It is important to emphasize the importance of everyone's collaboration in order for the action to be effective;

b) Creating partnerships: Establish partnerships with private skip companies, trade associations and residents together with the PMB. These partnerships will help provide financial resources, infrastructure and manpower;

c) Infrastructure and logistics: Determine which will be the strategic points for voluntary collection to deposit the CDW where the PEV (Voluntary Delivery Points) will be installed which are currently dump sites. It will be necessary to make buckets available at ground level for each type of waste according to the separation method established (brown and gray);

d) Training and qualification of waste collectors: Train the team responsible for the service, sharing proper waste separation practices, safe handling and the equipment used;

e) Dissemination and guidance: Maintain a schedule for disseminating details of the collection and the results already achieved. This action is important to keep the population engaged and aware of the benefits that the collection of MSW generates. This action can be carried out through information materials on social media and the distribution of leaflets, posters and banners at strategic locations in the city and at waste collection companies.

f) Gradual implementation: Start RSCD collection in a pilot area or specific neighborhood in the city. This will allow the system to be evaluated and adjusted before expanding it to other neighborhoods and eventually covering the entire city of Betim-MG;

g) Monitoring and evaluation: Establish PDCA (*Plan. Do. Check. Action*) monitoring mechanisms to track the efficiency and progress of DRSC collection. Carry out periodic analyses and collections to identify possible points for improvement and make the necessary adjustments;

h) Recycling and proper disposal at the plant: Ensure that the CSR waste collected is sent to the plant and that recycling is guaranteed in the correct processing of these materials; and

i) And finally, the theoretical potential for recyclable waste (PTGR) is the starting point for planning the municipal policy for recycling MSW, it indicates what is generated by the population within the municipality and defines the project's operational targets and the monitoring of the recovery rate (MINAS GERAIS, 2021).

7.6 RESULTS AND DISCUSSION

The theoretical potential for recyclable waste (PTGR) according to Eq . (1) is the starting point for planning the municipal policy for recycling the MSW indicates that is generated by the population within the municipality and according to IBGE (2022), the population of Betim has 411,846 inhabitants.

This defines the project's operational targets and the monitoring of the RSCD recovery rate. The calculation will be as follows with a 10-year forecast until 2033 with a permanent average of 18.91 kg/inhabitant/month and with a population increase of 0.2% p.a. according to the IBGE national average (2022) and starting in 2022 with 11% recycling of MSW from the plant currently in operation in 2023 PMB (2022) with a proposed increase of 5% p.a. for MSW recycling in 2033 reaching 60% of recycled MSW (if the project is implemented following the PDCA) (Table 1 and Graph 1).

The initial calculation: 411,846 inhab X 18.91 Kg/ inhab .month X 11% (recycling of the current RSCD in 2022) with a Population: 411,846 inhab (IBGE. 2022) and Production 18.91 Kg/hab.month (PMP. 2022 with an increase in population of 0.2% per year (Average according to IBGE statistics, 2022.

Table 2 - The theoretical potential for recyclable waste (PTGR) 2022 to 2023.

year	Population (inhab)	Population increase Betim-MG (0.2% p.a.)	Per capita generation Kg/inhab/ month (0.2 % p.a.)	Generation RS hab/month total (tons)	GPC Total annual per capita generation (tons)	Proposed 5% p.a. increase for CSR recycling (%)	PTGR - Theoretical Potential of Annual Recyclables (tons)	GPU- Generation Produced at the Plant 5% per year (tons)	DG - Annual waste generation deficit with recycling (tons)
2022	411.846		18.91	7.788,01	93.456	11	10.280	33.600	59.856
2023	411.846	412.670	18.95	7.803,58	93.643	15	14.046	35.280	58.363
2024	412.670	413.495	18.99	7.834,83	94.018	20	18.804	37.044	56.974
2025	413.495	414.322	19.02	7.866,20	94.394	25	23.599	38.896	55.498
2026	414.322	415.151	19.06	7.897,70	94.772	30	28.432	40.841	53.931
2027	415.151	415.981	19.10	7.929,32	95.152	35	33.303	42.883	52.269
2028	415.981	416.813	19.14	7.961,07	95.533	40	38.213	45.027	50.506
2029	416.813	417.647	19.18	7.992,94	95.915	45	43.162	47.279	48.636
2030	417.647	418.482	19.21	8.024,95	96.299	50	48.150	49.643	46.656
2031	418.482	419.319	19.25	8.057,08	96.685	55	53.177	52.125	44.560
2033	419.319	420.157	19.29	8.089,34	97.072	60	58.243	54.731	42.341

Source: Prepared by Diniz R.G.N (2023).

From the projected PTGR results, it was found that the initial total *per capita* generation was 93,456 t in 2022, ending in 2033 with 97,072 t.

The theoretical annual recyclable potential (PTGR) started at 10,240 t in 2023 and ended at 58,243 t in 20233.

In relation to the generation produced at the plant (GPU) in 2022 it produced 33,600 t, with an increase of 5% in the projection, ending 2033 with 54,731 t.

In relation to the waste generation deficit (DGR) in relation to the plant's recycling in 2022 it has a deficit of 59,856 t and in 2033 it will have a deficit of 42,341 t. The graph shows that in 2033 there will be an inversion of the theoretical recyclable potential in relation to the generation deficit (Figure 1).

It can then be concluded that if a projection is made to recycle 60% of the construction and demolition waste in the municipality of Betim-MG in 2033, the generation produced at the plant would be 54,731 tons and the theoretical annual recyclable potential would be 58,254 tons.

Figure 7 - GPC (Total annual *per capita* generation) *X* PTGR (Theoretical annual recyclables potential) *X* DG (Waste generation deficit in annual tons)

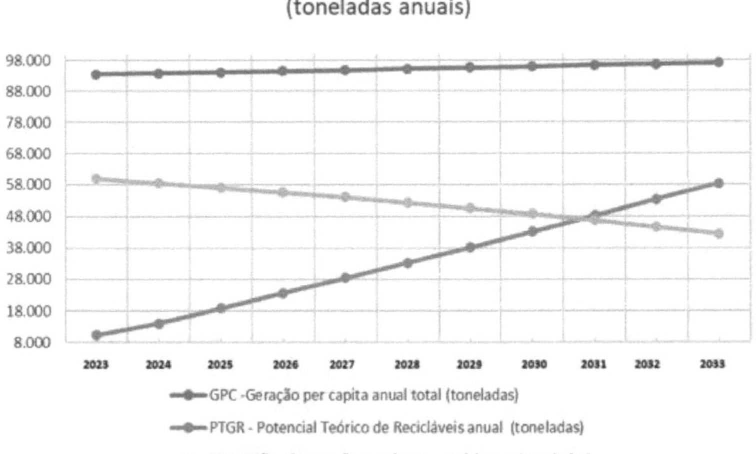

Source: Prepared by Diniz R.G.N (2023).

At the end of the proposed study, it is hoped to obtain a compilation of the management and recycling of solid waste in the city of Betim/MG by calculating the theoretical potential for recyclable waste (PTGR): according to the Implementation of Selective Collection in the Municipalities of Minas Gerais by the State Secretariat for the Environment and Sustainable Development. The target could be 60% in 10 years, with a 0.5% increase in recycling each year. The aim is to design and implement an efficient selective collection service that can serve as an example for other municipalities with similar characteristics.

From the data analyzed, it can be concluded that if a projection is made to recycle 60% of the CDW in the municipality of Betim-MG, with a projection for the year 2033, with 60% use of the generation produced in the municipality, the plant would recycle 54,731 tons and the theoretical annual recyclable potential would be 58,254 tons.

7.7 CONCLUSIONS

The need to implement the PTGR in parallel with the PDCA in order to have joint actions and projects to improve waste management in the city of Betim-MG. The deficits in the current actions of the Municipality of Betim-MG need to be changed and implemented.

The benefits of setting up a CSRD collection are numerous and, when properly implemented and coordinated, can produce positive results (Graph 1).

It is therefore important for municipalities to seek sustainable development solutions to ensure the collection, handling, reuse and correct disposal of CDW.

7.8 REFERENCES

ABRELPE - BRAZILIAN ASSOCIATION OF PUBLIC CLEANING AND SPECIAL WASTE. **Panorama of Solid Waste in Brazil 2022**. São Paulo: ABRELPE. 2022.

BRAZILIAN ASSOCIATION OF TECHNICAL STANDARDS. **NBR 10.004: Solid waste - Classification**. Rio de Janeiro. 2004.

BRAZIL. **Law No. 12.305 of August 2, 2010**. Establishes the National Solid Waste Policy; amends Law No. 9.605. of February 12, 1998; and makes other provisions. Federal Official Gazette. Brasília. August 2, 2010.

CEMPRE - CORPORATE COMMITMENT TO RECYCLING. **Municipal waste: integrated management manual**. 4. ed. São Paulo. SP. 316 p. CEMPRE. 2018.

BRAZILIAN INSTITUTE OF GEOGRAPHY AND STATISTICS - IBGE. **Betim**. IBGE. 2023. Available at: <https://cidades.ibge.gov.br/brasil/mg/betim/panorama>. Accessed on: 18 Nov 2023.

MINAS GERAIS. State Secretariat for the Environment and Sustainable Development **Guidelines manual: implementation of selective collection in the municipalities of Minas Gerais.** Belo Horizonte. 2021.

MIRANDA. Nathalia Mercedes; MATTOS. Ubirajara. Aluízio de Oliveira. Review of Models and Selective Collection Methodologies in Brazil. **Sociedade&Natureza.** [S. l.]. v. 30. n. 2. p. 1-22. 2018. DOI: 10.14393/SN-v30n2-2018-1.

PGRS/PTI - **Solid Waste Management Program 2019-2022**. Juliana Elisabete Correia. et al0. coordination Leilane Soares Pereira de Sousa. Sergio Angheben. 1.ed. Foz do Iguaçu.PR: FundaçãoPTI.2023.

SINIS - National Sanitation Information System. **Thematic Diagnosis of Urban Solid Waste Management 2021**. Brasília. 2022 Available at: <https://www.gov.br/mdr/pt-br/assuntos/saneamento/snis/produtos-do-snis/diagnosticos/diagnosticos_snis> . Accessed on: 15 Nov 2023.

8 CHAPTER V

TECHNICAL PRODUCT: TECHNICAL MANUAL FOR MANAGING PUBLIC RECYCLING PLANTS FOR CONSTRUCTION AND DEMOLITION WASTE

8.1 PRESENTATION

8.1.1 Technical Manual

This Technical Manual contains guidelines for the management of the Recycling Plant in the city of Betim, Minas Gerais, as well as rules and information on construction and demolition waste.

This PTT - Technical Technological Product is the result of research by the MPSTA - Professional Master's Degree in Sustainability and Environmental Technology of the Federal Institute - IFMG - Bambuí Fields/MG and was based on the project management tool developed by the PMI - Project Management Institute, the PMBOK - Project Management Body of Knowledge.

Project management is the application of knowledge, skills, tools and techniques to project activities in order to meet project requirements. Project management tools are applied globally and in a wide variety of sectors, thus guaranteeing the project's chances of success (PMI, 2008). According to Slack, Jones and Johnst (2018), every detail of the project must be analyzed to ensure that its objectives are met effectively.

O PMBOK é um guia de boas práticas em gerenciamento de projetos desenvolvido pelo PMI. Ele fornece um conjunto de processos, técnicas e áreas de conhecimento que são amplamente reconhecidos e utilizados por profissionais de gerenciamento de projetos em todo o mundo.

8.1.2 Recycling Plant in Betim MG

The recycling plant in Betim - a municipality in the metropolitan region of Minas Gerais - is managed by ECOS - Empresa de Construções, Obras, Serviços, Projetos, Transporte e Trânsito, a local authority in the city since 2018, where it now manages municipal public policies that were previously carried out by municipal departments and companies.

According to ECOS, the plant was acquired through a partnership with Banco do Brasil, is located in the Citrolândia neighborhood in Betim - MG and crushes 3,000t of rubble per month. It has a bidding contract with Via Solo, which manages the plant, according to ECOS. The plant has a built-up area of 18,000 m² and 2,200 m² of floor space.

The plant operates from 8am to 5pm from Monday to Friday, including rainy days. There is 24-hour surveillance, the operating structure has a guardhouse and night and emergency lighting, and the conditions of access roads, local vegetation and noise were taken into account when installing the plant.

According to the managers, the waste received at the plant is of Class A type. When the waste is received, transport controls are carried out and the raw materials are identified. When receiving the rubble, the MTR - Waste Transport Manifest - is required, a document issued for the transportation of solid waste by the State Environmental Foundation (FEAM), and the volume of waste disposed of at the plant is also identified. Also on receipt, the presence of contaminants or non-crushable waste such as earth, vegetation, etc. is analyzed. If there is an excessive amount of contaminants or if the waste does not conform to class A, the plant does not accept its disposal.

During pre-processing, visual, manual and magnetic sorting is carried out. During the processing process, granulometry is checked. The recycled material is packaged in the open air and is divided by type of granulate - gravel 0 or gravel 1.

The granulate is removed periodically according to the needs of Betim City Hall (PMB). The plant area is cordoned off and signposted by completely enclosing the area with a fence and access gate, access is controlled by a gate and guardhouse, there are safety signs with traffic and information signs, a green belt and distance from inhabited areas controls noise, while emissions and pollutants are controlled by water dispersion against dust.

8.1.3 Objective

The culture of consumerism and the unbridled generation of waste make it necessary to expand studies and actions on the problem, so this manual is experimental in nature and is not intended to be a definitive solution to the management of the object of the research.

The aim of this manual is to encourage the adoption of management techniques, providing practical guidelines for carrying out the project in such a way as to comply with legislation and guarantee the effectiveness of the management provided for in CONAMA 307/2002 and in the Municipal Waste Management Plan of the Municipality of Betim, and thus encourage the adoption of CDW - Construction and Demolition Waste - recycling techniques in the municipality, contributing to the fulfillment of Agenda 21 and cooperating with the preservation of the environment, guaranteeing a better life for future generations.

8.1.4 Regulatory instruments

Figure 1 below lists the main federal, state and municipal laws and regulations relating to solid construction waste.

Figure 1 - Legislation and regulations on the management of CSRD

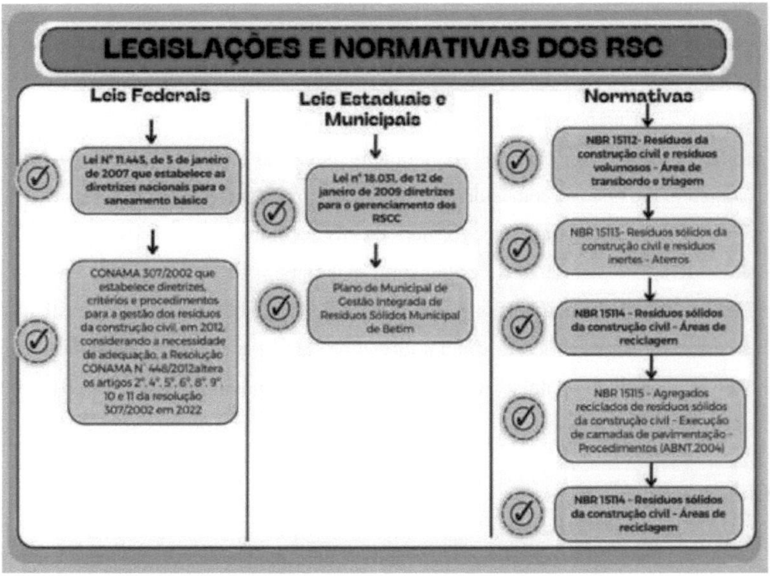

Source: Prepared by Diniz R. G. N, 2024.

8.1.5 Municipal Solid Waste Plan

In 2015, the municipality's Department of the Environment drew up t h e Municipal Plan for Integrated Solid Waste Management in Betim (PMGIRSB), following the guidelines of CONAMA Resolution 307 of July 5, 2002, from which Betim began to develop actions to minimize the impacts caused by solid waste.

According to the PMGIRSB (2015) and according to data collected in 2009, the management of solid construction and demolition waste (SWDW) in the municipality, like the majority of Brazilian cities, had not been providing adequate management of Construction and Demolition Waste/Construction Waste (SWDW/CWR), nor had it been following CONAMA resolution 307 and ABNT standards.

In this way, the PMGIRSB set targets for the reuse and recycling of CDW in 2015, to be achieved by 2020, and one of the goals of the municipal plan was to create municipal legislation for better management.

8.1.6 Waste classification

According to CONAMA 307/2002, waste is classified as A, B, C and D, as shown in Figure 2 below:

Figure 2 - Classification of CDW according to CONAMA 307/2002

Table 2: Classification of RSCDD according to CONAMA 307/2002

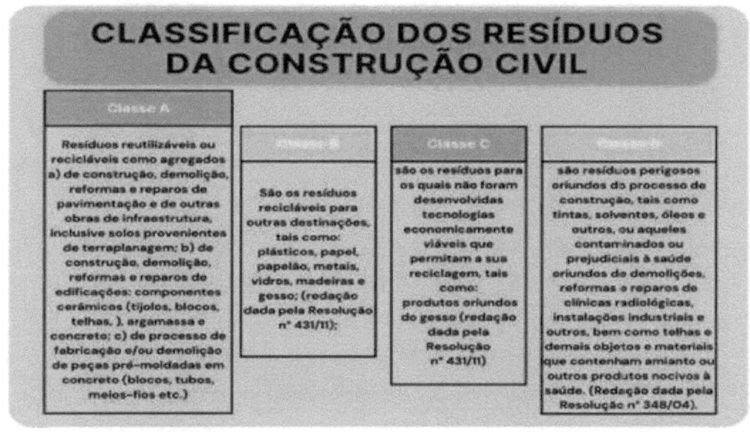

Source: Prepared by Diniz R. G. N, based on: BRASIL, 2002.

8.2 PLANT PROJECT PLANNING

8.2.1 Opening Term of the Plant

A project charter is a formal document that marks the start of a project. It is developed by the project manager or the team responsible and approved by stakeholders such as the mayor or municipal legislators. The project charter defines the project's context, objectives, stakeholders, key deliverables and constraints.

How to do it

The project manager, together with the team in charge, will draw up a document containing the information described below. Once it has been drawn up, approval will be sought from the interested parties, which could be the mayor, the environment secretary or the municipality's legislators.

Title: Opening Statement for the Betim CSR Recycling Plant Project

Project Name: CSR recycling plant

Start Date: (date the plant started operating)

Sponsor: Betim City Hall/ Partner companies (please quote)

Project Manager: Plant Manager (name)

Team: (Describe the team that will be working at the plant)

Project objective: (Describe the objective of the CSR recycling plant)

Project Justification: (Explain why the project is necessary, including expected benefits and problems to be solved)

Project output: (What types of crushing will be carried out at the Plant)

Project Restrictions: (List possible problems that may occur during management, such as lack of resources, etc.)

Key Stakeholders: (Describe all those involved in the project, everyone who will be "affected" in some way by the operation of the plant).

[Mayor] [ECOS Company] [Municipal Secretary for the Environment] ....

Preliminary Schedule: (Describe all the tasks that need to be carried out at the plant including important dates, milestones, etc.)

Preliminary Budget: (Describe the estimated cost of running the plant)

Initial Risks: (List the possible risks that may occur during the management of the plant process, as well as preliminary mitigation strategies)

Approvals: This Project Charter is approved by: (name(s))

Signature of Sponsor: (Date)

Signature of Project Manager: (Date)

FIGURE 3 - PLANT CONVEYOR BELT

Source: Prepared by Diniz R. G. N, 2023

8.3 PLANT PROJECT MANAGEMENT PLAN

A project management plan is a fundamental document that describes how the project will be executed, monitored, controlled and closed. It provides a framework to guide all project activities and is developed by the project manager in conjunction with the project team and relevant stakeholders. Here is a simplified example of how a project management plan can be structured.

How to do it

The project team, led by the manager, draws up a step-by-step document of the project process as described below;

Introduction: Draw up an introduction containing the plan's objective, the scope of the project and references),

Project Management Processes: Describe the processes below in as much detail as possible.

Project Integration: Describe how the plant's management processes will be coordinated and integrated.

Project Scope: Describe how the scope will be defined, validated and controlled.

Project Schedule: Defining how activities will be sequenced, estimated and controlled.

Project Cost: Describe how financial resources will be estimated, budgeted and controlled.

Project Quality: Describe the quality requirements of the plant's crushing process.

Project Human Resources: Describe how the plant team will be acquired, developed, managed and released.

Project Communications: Detail how process information in the plant will be collected, distributed, stored and managed.

Project Risk Management: Identify and analyze the risks to the plant's operation and describe response strategies.

Project Procurement: Describe how procurement will be planned, executed and managed.

Project Stakeholders: Describe how stakeholders will be identified, involved and managed throughout the operation of the plant.

Organizational Structure: Describe the plant's governance structure, including roles and responsibilities.

Project timetable: Describe the timetable for the plant's activities.

Project Budget: Describe the plant's budget, including cost estimates and allocation of financial resources.

Communication Plan: Describe the methods of communication between plant staff and stakeholders.

Risk Management Plan: Describe the main project risks and the associated mitigation strategies.

Change Controls: Describe the procedures for requesting, evaluating and implementing changes to the plant's process.

Project Closure: Describe the plant's acceptance criteria and the procedures for formally closing down its operation.

Approval: Describe at the end, this plan is approved by and take signature of Plant Manager and date.

Figure 4 below lists the stages of the plant's project management.

Figure 4 - Stages of the simplified power plant project management plan

.

PLANO DE GERENCIAMENTO DO PROJETO DA USINA

INTRODUÇÃO:

PROCESSOS DE GERENCIAMENTO DE PROJETOS

- Integração do Projeto
- Escopo do Projeto
- Cronograma do Projeto
- Custo do Projeto
- Qualidade do Projeto
- Recursos Humanos do Projeto
- Comunicações do Projeto
- Gerenciamento de Riscos do Projeto
- Aquisições do Projeto
- Partes Interessadas do Projeto
- Estrutura Organizacional
- Cronograma do Projeto
- Orçamento do Projeto
- Plano de Comunicação
- Plano de Gerenciamento de Riscos
- Controles de Mudança
- Encerramento do Projeto
- Aprovações:

INFORMAÇÕES IMPORTANTES

O esboço do projeto é importante para a visão geral do projeto.
Os prazos deverão ser respeitados.
A comunicação da equipe deve acontecer frequentemente.
Cada etapa do projeto necessita de avaliação e reflexão.

8.4

The project scope refers to the clear and detailed definition of the type of product that will be processed at the plant. It establishes the objectives, deliverables, activities, resources required, limits and acceptance criteria. The scope is essential to ensure that all parties involved understand what is and isn't included in the project, helping to avoid deviations, conflicts and communication problems.

How to do it

Manager and team, draw up a document entitled Project Scope Statement (PSD) or Project Plan containing the following information:

Objectives: Statement of the results the plant intends to achieve.

Deliveries: Specify which type of crushing the plant will deliver (gravel 0, 1...)

Requirements: Specify the plant's characteristics and performance criteria in detail.

Timetable: Describe the sequence of activities that will be practiced at the plant.

Budget: Describe all the financial resources needed to run the plant.

Risks: Identify and analyze potential events that could affect the plant's operation (e.g. rain, popular demonstrations...)

Constraints: describe all the Limitations that may impact on the operation of the plant, such as tight deadlines, limited resources, legal restrictions, etc.

Acceptance criteria: Clear criteria that define when a delivery is considered completed and satisfactory. (This can be a form given to the PMB works department.

It is important that the project scope is documented in a clear and concise manner. This document serves as a reference throughout the execution of the project, helping to maintain focus on the established goals and facilitating the management of stakeholder expectations.

8.5 ACTIVITIES

Project activities determine the specific tasks that need to be carried out in order to achieve the objectives and deliver the recycled aggregates in accordance with the scope of the project, taking into account the deadlines and budgets set. They are essential for turning project planning into action and are organized in a logical sequence to ensure the successful completion of the project.

How to do it

This task is assigned to the plant manager, who will describe the sequence of activities according to the plant's reality, but with the minimum requirements listed below:

Initial planning: Hold meetings with the plant management team, establish teams and assign responsibilities.

Requirements gathering: Hold meetings with those involved in the plant, the mayor, legislators and secretaries, document the characteristics of the recycled aggregate.

Analysis: Carry out feasibility analysis of recycled aggregate.

Describe the activities and schedules: Describe the plant's activities, from receiving the rubble to delivering the recycled aggregate, establish daily production quantities, describe equipment downtime for preventive maintenance, corrective maintenance, describe idle time of machinery and equipment (consider hour/man/machine), draw up a schedule table of plant activities and share it with everyone involved.

Monitoring and Control: Tracking the project's progress against the schedule, managing risks and problems, ensuring compliance with the requirements set out in the Project Management plan.

Closure: (consider - in this case - closure, the end of the bidding contract), document lessons learned, hold meetings to review experience in running the plant, archive plant documentation. Figure 5 below shows a model of the plant's operating flow, inspired by the VSM.

Figure 05 - Plant operation flow model

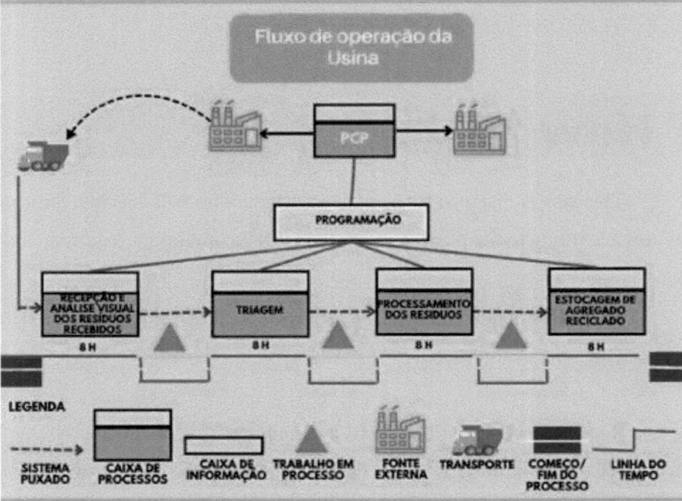

Source: Prepared by Diniz R. G. N, 2024, based on *Lean Manufacture* Value Stream Map, 2024

These generic activities can be adapted according to the plant's needs. It is important that the activities are clearly defined, assigned to responsible team members and closely monitored by management to ensure effective management.

Figure 06 - Demonstration of plant granulometry - Crushed stone 2

Source: Diniz R. G. N, 2023.

8.6 COST MANAGEMENT

Cost management is an essential part of project management, which involves planning, estimating, controlling and monitoring the financial resources required to carry out a project within the established budget. Here are some key steps in the cost management process:

How to do it

The project manager, together with the planning team, will draw up the cost management plan as described below.

Cost Planning: Establish the budget, based on the requirements, scope and resources needed to draw up the tender, identify and estimate the costs of all the plant's activities, including labor, materials, equipment, contracted services and any other costs not mentioned in this manual, develop a Cost Analysis Structure (CAS), which divides the project's costs into specific categories for better understanding and control.

Cost Estimating: Prepare accurate estimates of the costs associated with each plant activity, using technical tools such as analog estimating, consider the risks and uncertainties involved in cost estimates and add a reserve to deal with unforeseen events.

Budgeting: Allocating the estimated costs for each plant activity over time, creating a detailed budget that shows how the financial resources will be used over the period of the tender contract, ensuring that the budget is consistent with the processes practiced at the plant so as not to burden the public budget, developing a cost baseline, which serves as a reference for monitoring and controlling actual performance in relation to plan.

Cost Control: Create reports to regularly monitor costs, comparing actual figures with the planned budget, identify and analyze cost variations, investigating the underlying causes and implementing corrective measures when necessary, hold monthly meetings to analyze cost reports and make decisions when necessary.

Performance Reports: Prepare financial performance reports that clearly communicate the current status of the plant's costs, highlighting significant variations, trends and forecasts for the future. Figure 7 below lists the Cost Management Cycle.

Figure 07 - Cost management cycle

Source: Prepared by Diniz R.G.N

TIP:

An important tip for cost management is to adopt a proactive and systematic approach to identifying, monitoring and controlling costs over time

8.7 RESOURCE MANAGEMENT

Resource management involves identifying, distributing, planning, controlling and optimizing the resources needed to carry out project activities effectively. These resources can include people, equipment, materials and finance.

How to do it

The plant manager, together with the planning team, identifies all the resources needed for the plant's process to run smoothly, taking into account people, equipment, materials and finance, as specified below.

Resource identification: Identify all the resources needed to improve the plant's processes, taking into account the scope, requirements and objectives of the project carried out previously. To do this, human resources, equipment, tools and financial resources must be considered.

Resource Allocation: Allocating the listed resources efficiently, assigning the right people to the right tasks and ensuring that material and financial resources are available when needed. Take into account the skills, availability and workload of team members when assigning tasks.

Resource Planning: Develop a plan that details how resources will be used in the plant process, this can include a schedule of activities that shows when resources will be needed.

Resource Control: Monitoring the use of resources in the plant process, to ensure that they are being used efficiently and effectively. This can involve tracking the time spent by team members on each activity, monitoring the use of equipment and materials and controlling the plant's budget.

Resource Optimization: Analyze and Identify opportunities to optimize the use of resources, such as reallocating team members to areas where they are most needed, using equipment more efficiently or finding ways to reduce costs without compromising the quality of the process in the plant. Periodically reassess resource plans.

Resource Conflict Management: Resolve conflicts that may arise in relation to the availability or use of resources, such as overlapping schedules for team members or disputes over priorities for the use of equipment. Communicate effectively with all stakeholders involved to find solutions that meet the needs of the project.

TIP:

The VSM model can be used to control and propose improvements

Figure 8 - Recycling plant in Betim

Source: Diniz R. G. N, 2023.

8.8 COMMUNICATION

Project communication, as defined by the PMBOK, follows a structured pattern to ensure that all stakeholders receive relevant information in a timely and accurate manner for the proper performance of project processes.

How to do it

Planning team and manager, structure report, internal communication, letters and documents, as per example below, hold meetings with all those involved in the management of the plant to make decisions periodically.

Stakeholder communication document template:

Subject: Update on the Betim recycling plant project

Dear team members and others involved,

This is an update on the recycling plant in Betim

Introduction: A brief background to the plant and its importance to society.

Project Objectives: Recap of the main objectives of the project as defined in the project plan.

Project Status: Current status of the plant

Scope: Description of the current scope and any significant changes that have occurred.

Costs: Current state of the plant's budget, including spending to date and forecasts of future costs.

Quality: Discussion of the quality of the work done so far and any quality problems identified.

Risks: Update on the main risks identified and any mitigation measures implemented to keep the plant running.

Decisions and Actions Taken: Summary of the main decisions taken during the plant's operation.

Next Steps: Description of the plant's next management steps, including planned activities, assigned responsibilities and expected results.

Next Meeting: Details of the next meeting, including date, time and preliminary agenda.

Feedback: Request for feedback on project communication and any suggestions for improvement.

Thank you for your attention and I am available to answer any questions or concerns you may have.

Sincerely, [Your Name] [Your Position] [Date]

Figure 9- The plant's rubble reception area

Source: Diniz R. G. N, 2023.

8.9 ACQUISITIONS

In the context of project management, procurement refers to the acquisition of products, services or external resources that are necessary to ensure the success of the project. Managing acquisitions effectively is essential to ensure that the project has access to the necessary resources.

How to do it

Plant manager together with planning team, carry out procurement definitions, describe and control all the resources required for the process as described below:

Procurement Planning: Identify which products, services or resources will be procured externally. Develop a procurement plan that details the procurement needs, supplier selection criteria, appropriate procurement methods and expected delivery times. Determine the legal, regulatory and organizational requirements that must be followed during the procurement process.

Supplier selection: Evaluate potential suppliers based on criteria such as experience, reputation, technical capacity, financial capacity and performance history. Carry out tenders with the selected suppliers to obtain competitive bids and guarantee the best value.

Contracting and Contract Administration: Develop clear and comprehensive contracts that establish the terms and conditions of the acquisition, including scope, cost, deadlines, responsibilities and guarantees. Manage the performance of suppliers throughout the life cycle of the contract, monitoring quality, compliance with deadlines and the budget.

Closing contracts: Evaluate supplier performance at the end of the contract. Formalize the closure of the contract, including the release of any final payments due and the resolution of any outstanding issues.

8.10 REFERENCE

BRAZILIAN ASSOCIATION OF TECHNICAL STANDARDS. NBR 10.004 - **Solid waste** - Classification. 2004.

BRAZILIAN ASSOCIATION OF TECHNICAL STANDARDS. NBR 15112- **Resíduos da construção civil e resíduos volumosos** - Área de transbordo e triagem - Diretrizes para projeto, implantação e operação. 2004.

BRAZILIAN ASSOCIATION OF TECHNICAL STANDARDS. NBR 15113- **Solid waste from civil construction and inert waste** - Landfills - Guidelines for design, implementation and operation. 2004.

BRAZILIAN ASSOCIATION OF TECHNICAL STANDARDS. NBR 15114 - **Solid construction waste** - Recycling areas - Guidelines for design, implementation and operation. 2004.

BRAZILIAN ASSOCIATION OF TECHNICAL STANDARDS. NBR 15115 - **Recycled aggregates from solid construction waste** - Paving layers - Procedures. 2004.

BRAZILIAN ASSOCIATION OF TECHNICAL STANDARDS. NBR 15116 - **Recycled aggregates from solid construction waste** - Use in paving and concrete preparation without structural function - Requirements. 2004

BRAZIL. **CONAMA Resolution No. 348 of August 16, 2004**. Amends CONAMA Resolution 307, of July 5, 2002, including asbestos in the class of hazardous waste. Brasília, 2004.

Office of the Chief of Staff. Law No. 12.305, of August 2, 2010. **Establishes the National Solid Waste Policy; amends Law No. 9.605, of February 12, 1998**; and makes other provisions. Brasilia: DOU, 2010.

BRAZIL. **CONAMA Resolution No. 313 of October 29, 2002**. Provides for the National Inventory of Solid Industrial Waste. Federal Gazette: section 1, Brasília, DF, 2002, October 29, 2022.

JONES, A.; JOHNST, Robert. **Production management**. 8.ed. . v. 1. São Paulo, SP: Atlas, 2018

PMI-PROJECT MANAGEMENT INSTITUTE, INC (United States of America). **A Guide to Project Management Knowledge**: PMBOK Guide. 4. ed. rev. United States of America: PMI book Service Center, 2008. ISBN 978-1-933890-70-8SLACK, N.;

SEMMAD -Secretaria **municipal de meio ambiente e desenvolvimento sustentável**. PMB - Betim City Hall (Minas Gerais). 2015. Betim -MG, p. 1-128, 2015.

SHOU, W.; WANG, J.; WU, P.; WANG, X.; CHONG, H-Y. **A cross-sector review on the use of Value Stream Mapping. International Journal of Production Research**, v. 55, n. 13, p. 3906-3928, 2017.

9 CONCLUSION

Given the need to reduce the environmental impact of construction, it is essential to explore and implement options that promote efficient management of the waste generated. This waste has significant theoretical potential, offering opportunities for recycling and reuse.

The current scenario for construction waste recycling plants varies in different aspects. Although they face financial challenges, such as high operating costs, there are opportunities for sustainable investment, especially with incentive policies and growing awareness of the importance of recycling. Environmentally, these plants have the potential to substantially reduce the amount of waste sent to landfills, contributing to the conservation of natural resources and the reduction of pollution. Socially, their operation can generate local jobs and promote the sustainable development of communities.

Despite the existing potential, it is essential to improve the management of plants by investing in more efficient technologies and sustainable management practices. In addition, it is necessary to promote public policies that encourage recycling and the use of recycled materials in construction, as well as raising awareness among the various players involved about the importance of reducing, reusing and recycling waste. With integrated and collaborative approaches, it is possible to turn challenges into opportunities and move towards a more sustainable and resilient future for the construction sector.

10 REFERENCES

AUTODESK (USA). Revit: **BIM software for designers, builders and developers. In: Revit: BIM software for designers, builders and developers. 1.0-RC2**. [S. l.], 2022. Available at: https://www.autodesk.com.br/products/revit/overview?term=3-YEAR&tab=subscription. Accessed on: July 25, 2022.

BRAZIL - CHAMBER OF DEPUTIES. Ministry of Foreign Affairs, Environment Division, with the collaboration of the United Nations Environment Program, UNEP. **Parliamentary Action, 56, 1995. 56. United Nations Conference on Environment and Development,** Brasília-DF: Documentation and Information Center, Publications Coordination, v. 1, p. 472, 1995.

BRAZIL - National Secretariat for Environmental Sanitation. Network for training and technological extension in environmental sanitation. ReCesa, 2007. **Resíduos sólidos: PGRSU -Plano de gestão de resíduos Sólidos urbanos**: : guia do profissional em treinamento, Belo Horizonte- MG: Artes Gráficas Formato LTDA, v. 1, p. 100, 2007.

BRITO, D. M. de; FERREIRA, E. de A. M. **Evaluation of strategies for representation and analysis of construction planning and control using 4D BIM models**. Ambiente Construído, Porto Alegre, v. 15, n. 4, p. 203-223, Oct./Dec. 2015.DOI: http://dx.doi.org/10.1590/s1678- 86212015000400047.

BETIM CITY COUNCIL. Ordinary Law No. 6381, of June 29, 2018. **Provides for the transformation of the Betim construction, works, services, projects, transportation and traffic company** - ECOS INTO A MUNICIPAL PUBLIC AUTHORITY, AND GIVES OTHER PROVISIONS. Ordinary Law No. 6.381/2018, Betim, 2018

Cardoso, A. da C. F., Galatto, S. L., & Guadagnin, M. R. (2014). **Estimation of Construction Waste Generation and Feasibility Study of a Sorting and Recycling Plant**. Brazilian Journal of Environmental Sciences (RBCIAMB), (31), 1-10. Available at: https://www.rbciamb.com.br/Publicacoes_RBCIAMB/article/view/254/209 Last accessed on: Dec. 5, 2023

CATELANI, Wilt Silva (Brazil). Cbic - Brazilian Chamber of the Construction Industry. **Coletânea Implementação do BIM Para Construtoras e Incorporadoras: Fundamentos BIM** - Accessed on: Apr. 12, 2017.

CAVALCANTI, Francisco, P.; SILVEIRA, Jarbas, N. **Fundamentals of Project Management**. São Paulo: Atlas, 2016. Available at: https://doity.com.br/media/doity/submissoes/artigo-809f7a66f61a623503d615654422c5e23f884d52-arquivo.docx Last accessed on: 5 Dec. 23

CONAMA RESOLUTION NO. 307, OF JULY 5, 2002. **Establishes guidelines, criteria and procedures for the management of civil construction waste**. Available at: https://cetesb.sp.gov.br/licenciamento/documentos/2002_Res_CONAMA_307.pdf Last accessed on: Dec. 5, 2023

Contemporânea - Revista de Ética e Filosofia Política, v. 3, n. 3, 2023. ISSN 2447-0961 **Empresa de Construções, Obras, Serviços, Projetos, Transporte e Trânsito Annual newsletter**, 2020. Betim City Hall (Minas Gerais). Betim -MG, 2020 - Empresa de Construções, Obras, Serviços, Projetos, Transporte e Trânsito Annual report, 2022. Betim City Hall (Minas Gerais). Betim -MG, 2022.

CRUZ, N. F.; MARQUES, R. C. **Revisiting the determinants of local government performance**. Omega, v. 44, p. 91-103, Apr. 2014. Available at: https://www.sciencedirect.com/science/article/abs/pii/S0305048313000923?via%3Dihub Last accessed on: Dec. 5, 23

ECOS - Empresa de Construções, Obras, Serviços, Projetos, Transporte e Trânsito **Annual report, 2019**. Betim City Hall (Minas Gerais). Betim -MG, 2019.

FRASSON, S.. **Rubble recycling plants as agents in the valuation of waste generated by civil construction**, [s. L.], 12017.

GNECCO, verônica. **BIM for waste management in civil construction: a case study on the Cras project in the municipality of Biguaçu/SC.** 2019. 150 f. TCC (Bachelor's Degree) - Technology Center of the Federal University of Santa Catarina, Santa Catarina, 2018.

IBGE - Brazilian Institute of Geography and Statistics. Cities and States - **Betim.** Available at: https://www.ibge.gov.br/cidades-e-estados/mg/betim.html Last accessed on: Dec. 5, 2023

IBGE - Brazilian Institute of Geography and Statistics. **Population 1980**. In: Federal Government. IBGE. Population. [S. l.], 2021. Total population - 1980 - 2010. Available at:

LARUCCIA, **Mauro. Sustainability and Environmental Impacts of Construction**. Article, [s. l.], 2014.

MARCONI, Marina d; LAKATOS, Eva. **Fundamentals of scientific methodology**. 8. ed. actual. São Paulo: Atlas, 2017. ISBN 978-85-970-1076-3.

MY Municipality. In: **Data from Betim**. 1.0-RC2. [S. l.], 2015. Available at: https://meumunicipio.org.br/perfil-municipio/3106705-BetimMG?exercicio=2015. Accessed on: June 14, 2022.

NUNES, Jéssica Martins; LONGO, Orlando Celso; ALCOFORADO, Luciane Ferreira; PINTO, Gustavo Oliveira. **The Civil Construction sector in Brazil and the current economic crisis**. Available at: https://rsdjournal.org/index.php/rsd/article/view/7274/6553 Last accessed on: Dec. 5, 2023

PMGIRSB - **Betim municipal integrated solid waste management plan**. PMB - Betim City Hall (Minas Gerais). 2015, Betim -MG, p. 1-128, 2015.

PMI. **A Guide to Project Management Body of Knowledge (PMBOK® Guide)**-Fourth Edition. Available at: https://www.cin.ufpe.br/~if717/slides/PMBOK.pdf Last accessed on: 5 Dec. 23

BETIM CITY HALL. Law No. 6412, of September 12, 2018. **Provides for the municipal basic sanitation policy and makes other provisions.** Betim Minas Gerais, September 12, 2018.

Presidency of the Republic General Secretariat. **LAW NO. 12.965, OF APRIL 23, 2014**. Available at: https://www.planalto.gov.br/ccivil_03/_ato2011-2014/2014/lei/l12965.htm Last accessed on: Dec. 5, 2023

Ribas JR, Vicente TV dos S, Altaf JG, Troccoli IR. **Integration of actions in sustainable management**. REAd Rev eletrôn adm (Porto Alegre) [Internet]. 2017May;23(2):31-57. Available from: https://doi.org/10.1590/1413.2311.112.58086. Last accessed on: Dec. 5, 23

SEMMAD - **Municipal Secretary for the Environment and Sustainable Development.** PMB - Betim City Hall (Minas Gerais). 2015, Betim - MG, p. 1-128, 2015.

ZANETTI1, Gabriela Candeo; MARTINES, Karen Staquecini; TOSELLO, Maria Eunice Carvalho Tosello; SIMÕES, Rebeca Delatore. **Solid construction waste**. Encontro Nacional de Ensino, Pesquisa e Extensão, Presidente Prudente, ano 2015, v. 12, ed. especial, p. 350-358, 22 oct. 2015.

11 ANNEXES

TERMO DE COOPERAÇÃO QUE ENTRE SI CELEBRAM A EMPRESA DE CONSTRUÇÕES, OBRAS, SERVIÇOS, PROJETOS, TRANSPORTES E TRÂNSITO DE BETIM – ECOS E A SRA. RAVENA GLICERIA NOLL DINIZ, NA FORMA ABAIXO.

PARTÍCIPES:

EMPRESA DE CONSTRUÇÕES, OBRAS, SERVIÇOS, PROJETOS, TRANSPORTES E TRÂNSITO DE BETIM – ECOS, pessoa jurídica de direito público interno, autarquia integrante da Administração Pública Indireta do Município de Betim, constituída nos termos da Lei Municipal nº 6.381, de 29 de junho de 2018, inscrita no CNPJ sob o nº 74.077.025/0001-20, com sede no Centro Administrativo Papa João Paulo II, localizado na Rua Pará de Minas, nº 640, Bairro Brasileia, Betim/MG, CEP nº 32.600-412, neste ato representada por sua Presidente, Sr.ª Marinesia Dias da Costa Makatsuru, inscrita no RG sob o nº MG-3.900.672 e no CPF sob o nº 696.291.456-34, daqui por diante designado **1º PROPONENTE**.

RAVENA GLICERIA NOLL DINIZ, pessoa física, inscrita no RG sob o nº MG-12.698.238 e no CPF sob o nº 014.367.016-60, residente e domiciliada na Rua Clotilde Borges, nº 105, Bairro Jardim da Cidade, Betim/MG, CEP nº 32.604-288, daqui por diante designado **2º PROPONENTE**.

As partes acima identificadas ajustaram e, por este instrumento, celebram um Termo de Cooperação, em conformidade com as normas legais vigentes, mediante as seguintes cláusulas e condições:

CLÁUSULA PRIMEIRA – DO OBJETO

O presente Termo de Cooperação tem por objeto estabelecer uma parceria entre as partes para que a **2º PROPONENTE** possa analisar a Gestão de Resíduos Sólidos da Construção Civil e avaliar a viabilidade econômica e socioambiental da usina de reciclagem de material de construção na Cidade de Betim/MG, através da coleta de dados junto à **1º PROPONENTE**.

CLÁUSULA SEGUNDA – DO VALOR DO CONVÊNIO

Não haverá nenhum desembolso financeiro entre os partícipes, devendo cada um arcar somente com as despesas e custos internos que se fizerem necessários às participações e ao fiel cumprimento do Termo de Cooperação.

CLÁUSULA TERCEIRA – DAS OBRIGAÇÕES DOS PARTÍCIPES

3.1 – Competirá à 1º PROPONENTE:

3.1.1 – permitir que a **2º PROPONENTE** acesse a Usina de RCC localizada na região de Citrolândia;

PREFEITURA DE BETIM WWW.BETIM.MG.GOV.BR

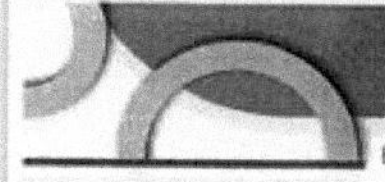
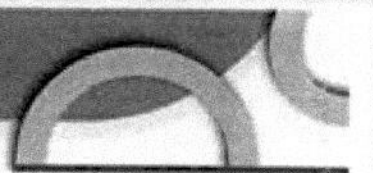

3.1.2 – fornecer as informações e os dados necessários para instruir adequadamente a pesquisa, desde que não sejam de cunho pessoal e/ou sigiloso;

3.1.3 – permitir que a **1ª PROPONENTE** acompanhe as atividades realizadas na Usina de RCC.

3.2 – Competirá à 2ª PROPONENTE:

3.2.1 – realizar a pesquisa, através das visitas técnicas e das coletas de dados, respeitando estritamente as orientações da **1ª PROPONENTE**;

3.2.2 – entregar à **1ª PROPONENTE**, ao final do prazo de vigência do Termo de Cooperação, o manual elaborado durante a pesquisa, a dissertação aprovada com os resultados da pesquisa e a cópia da aprovação da discente no mestrado;

3.2.3 – apresentar à **1ª PROPONENTE**, ao final do prazo de vigência do Termo de Cooperação, os resultados finais da pesquisa.

CLÁUSULA QUARTA – DOS GESTORES DO TERMO DE COOPERAÇÃO

A gestão do Termo de Cooperação caberá ao Diretor da Diretoria de Serviços Ambientais e ao Chefe do Setor de Resíduos, ambos os órgãos integrantes da estrutura orgânica da **1ª PROPONENTE**.

CLÁUSULA QUINTA – DO PRAZO DE VIGÊNCIA E DAS ALTERAÇÕES

5.1 – O presente Termo de Cooperação terá início com a assinatura deste instrumento e vigorará até o dia 31 de dezembro de 2023, podendo ser prorrogado, mediante acordo entre as partes, através de termo aditivo.

5.2 – As modificações do texto original poderão ser operadas por termos aditivos, com validade imediata após a assinatura do termo apêndice.

CLÁUSULA SEXTA – DA DENÚNCIA E DA RESCISÃO

6.1 – A denúncia ou a rescisão deste Termo poderá ocorrer a qualquer tempo, por iniciativa de qualquer um dos partícipes, mediante notificação com antecedência mínima de 30 (trinta) dias.

CLÁUSULA SÉTIMA – DA PUBLICAÇÃO

7.1 – O extrato do presente Termo de Cooperação será encaminhado para publicação no Órgão Oficial do Município de Betim até o 5º (quinto) dia útil do mês subsequente ao de sua assinatura, devendo ocorrer no prazo de até 20

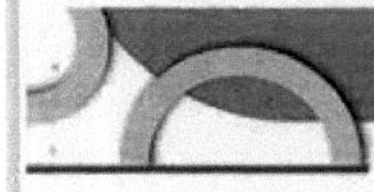

ECOS

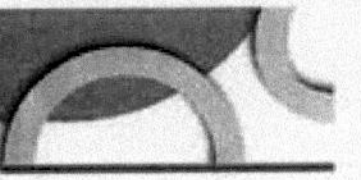

Empresa de Construções, Obras, Serviços, Projetos, Transportes e Trânsito de Betim
Rua Pará de Minas, 640, Brasiléia | Betim | MG - CEP.: 32600-412

(vinte) dias daquela data, em conformidade com o que estabelece o parágrafo único, do art. 61, da Lei Federal nº 8.666/1993.

CLÁUSULA OITAVA – DO FORO

8.1 – Fica eleito o foro da Comarca de Betim para dirimir qualquer dúvida ou litígio que porventura possa surgir da execução deste Termo de Cooperação, com expressa renúncia de qualquer outro, por mais privilegiado que seja.

E, por estarem justos e acordados, firmam este Termo em 02 (duas) vias de igual teor e forma, comprometendo-se a cumprir e a fazer cumprir, por si e por seus sucessores, em juízo ou fora dele, tão fielmente como nele se contém na presença das testemunhas abaixo, para que produza os devidos e legais efeitos.

Betim/MG, 11 de fevereiro de 2022.

Marinésia Dias da Costa Makatsuru
Presidente da ECOS
(1º PROPONENTE)

Raphael Martins Filho
Diretor Jurídico da ECOS
(1º PROPONENTE)

Ravena Glicéria Noll Diniz
(2º PROPONENTE)

TESTEMUNHA 1
Assinatura:
Nome:
CPF: 953.357.436-53

TESTEMUNHA 2
Assinatura:
Nome: Francimeyre da Silva Ribeiro
CPF: 95339957691

PREFEITURA DE BETIM WWW.BETIM.MG.GOV.BR

Contemporânea Revista de Ética e Filosofia Política

DECLARAÇÃO

A Contemporânea Revista de Ética e Filosofia Política, ISSN 2447-0961, declara para os devidos fins, que o artigo intitulado **"ANÁLISE DO GERENCIAMENTO DE RESÍDUOS SÓLIDOS DE CONSTRUÇÃO E DEMOLIÇÃO (RSCD), NO MUNICÍPIO DE BETIM - MG"** de autoria de **Ravena Glicéria Noll Diniz, Carlos Fernando Lemos, Fernanda Carla Wasner Vasconcelos, Ariane Flávia do Nascimento**, foi publicado no v. 3, n. 3, p. 2428-2441.

A revista é on-line, e os artigos podem ser encontrados ao acessar o link:

https://revistacontemporanea.com/ojs/index.php/home/issue/view/13

DOI: https://doi.org/10.56083/RCV3N3-067

Por ser a expressão da verdade, firmamos a presente declaração.

São José dos Pinhais, 21 de Março de 2023.

QR de validade da publicação

Prof. MSc. João Paulo Perbiche
Editor-chefe

CERTIFICADO

TEMÁTICA 2022
PLANETA TERRA, ÁGUA E AR
consciência, conservação e educação

Certificamos que

o trabalho intitulado "ANÁLISE DO GERENCIAMENTO DE RESÍDUOS SÓLIDOS DE CONSTRUÇÃO E DEMOLIÇÃO (RSCD), NO MUNICÍPIO DE BETIM - MG" de autoria Ravena Glicéria Noll Diniz, Ariane Flávia do Nascimento, Fernanda Carla Wasner Vasconcelos, Amanda Ferreira França Ribeiro, Carlos Fernando Lemos, foi eleito um dos melhores trabalhos do XIX Congresso Nacional de Meio Ambiente de Poços de Caldas, realizado no período de 21 a 23 de setembro de 2022 em Poços de Caldas, Minas Gerais, Brasil.

Prof. Dr. Fabrício dos Santos Ritá
Membro da Comissão
Técnico Científica

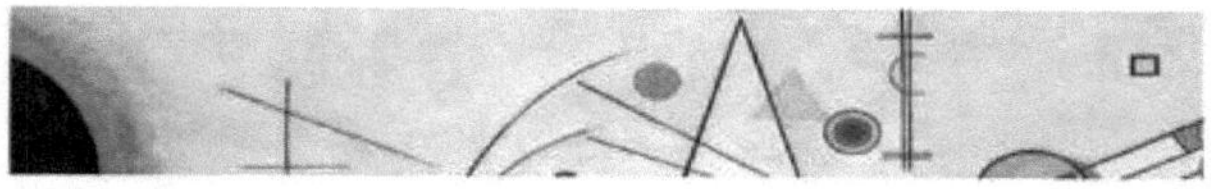

Contemporânea
Contemporary Journal
3(3): 2428-2441, 2023
ISSN: 2447-0961

Artigo

ANÁLISE DO GERENCIAMENTO DE RESÍDUOS SÓLIDOS DE CONSTRUÇÃO E DEMOLIÇÃO (RSCD), NO MUNICÍPIO DE BETIM - MG

ANALYSIS OF THE MANAGEMENT OF SOLID CONSTRUCTION AND DEMOLITION WASTES (HSW) IN THE CITY OF BETIM – MG

DOI: 10.56083/RCV3N3-067
Recebimento do original: 24/02/2023
Aceitação para publicação: 22/03/2023

Ravena Glicéria Noll Diniz
Mestranda Profissional em Sustentabilidade e Tecnologia Ambiental
Instituição: Instituto Federal de Ciência e Tecnologia Minas Gerais (IFMG) - Campus Bambuí, Pontifícia Universidade Católica (PUC-MG)
Endereço: Rodovia Bambuí, Medeiros, Km 05, Bambuí - MG, CEP: 38900-000
E-mail: ravena.diniz@ufv.br

Carlos Fernando Lemos
Doutor em Geoquímica Ambiental
Instituição: Universidade Federal de Viçosa (UFV) - Florestal
Endereço: Rodovia LMG 818, Km 6, CEP: 35690-000, Florestal - MG
E-mail: fernando.lemos@ufv.br

Fernanda Carla Wasner Vasconcelos
Doutora em Ciências
Instituição: Instituto Federal de Ciência e Tecnologia Minas Gerais (IFMG) - Campus Bambuí, Pontifícia Universidade Católica (PUC-MG)
Endereço: Rodovia Bambuí, Medeiros, Km 05, Bambuí - MG, CEP: 38900-000
E-mail: ravena.diniz@ufv.br

Ariane Flávia do Nascimento
Doutora em Ciência Animal
Instituição: Instituto Federal de Ciência e Tecnologia Minas Gerais (IFMG) - Campus Bambuí, Pontifícia Universidade Católica (PUC-MG)
Endereço: Rodovia Bambuí, Medeiros, Km 05, Bambuí - MG, CEP: 38900-000
E-mail: ariane.nascimento@ifmg.edu.br

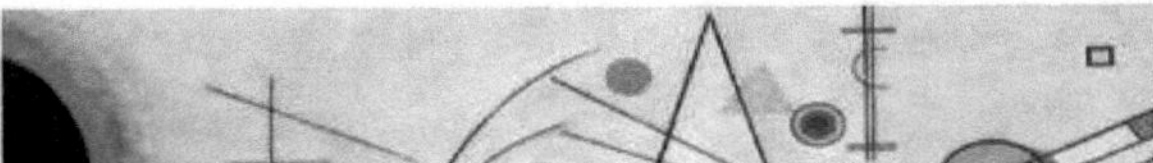

RESUMO: A Eco-92 estabeleceu políticas de responsabilidade ambiental, em uma agenda que pouco se avançou nos últimos trinta anos. Portanto, é necessário intensificar os nossos esforços para coletar, analisar e discutir informações sobre a gestão dos resíduos sólidos de construção civil e de demolição (RSCD). Este artigo teve por objetivo diagnosticar a gestão dos RSCD, em Betim. A Associação Brasileira de Empresas de Limpeza Pública e Resíduos Especiais destaca a região Sudeste como maior geradora de resíduos sólidos da Construção e demolição do Brasil, representando aproximadamente 50 por cento do total gerado no país. Este trabalho, utilizou o método de pesquisa aplicada, de natureza descritiva, a partir de uma abordagem qualitativa, por meio de um estudo de caso. Os dados foram obtidos através de documentos internos de controle de produção da Empresa ECOS e do Plano de Gerenciamento Integrado de Resíduos Sólidos do Município de Betim (MG), no período de 2015 à 2022. Foram analisados os dados da geração de resíduos em 2015 e posterior a geração e reciclagem desde o funcionamento da Usina de Reciclagem de Betim em 2018 a 2022. Constataram como resultados desta pesquisa que o município propôs o Plano Municipal de Gestão Integrada de Resíduos Sólidos Municipal, em 2015, conforme a legislação vigente, com metas que estão sendo cumpridas, desde 2018. Assim, nos últimos 5 anos, reciclou aproximadamente 150 000 toneladas, transformando o agregado reciclado em cobertura de estradas rurais, fazendo valer o cumprimento da agenda 2030, preservando o ecossistema e tornando-se cidade modelo para outros municípios.

PALAVRAS-CHAVE: Sustentabilidade, Agenda 2030, Responsabilidade Ambiental, Plano Municipal de Gestão Integrada de Resíduos Sólidos.

ABSTRACT: Eco-92 established policies for environmental responsibility, in an agenda that has advanced little in the last thirty years. Therefore, it is necessary to intensify our efforts to collect, analyze and discuss information about the management of solid construction and demolition waste (CSWR). This article aimed to diagnose the management of CSWR, in Betim. The Brazilian Association of Companies for Public Cleaning and Special Waste highlights the Southeast region as the largest generator of solid waste from construction and demolition in Brazil, representing approximately 50 percent of the total generated in the country. This work used the method of applied research, of a descriptive nature, from a qualitative approach, through a case study. The data were obtained through internal production control documents of the ECOS Company and the Integrated Solid Waste Management Plan of the Municipality of Betim (MG), for the period 2015 to 2022. The data of waste generation in 2015 and subsequent generation and recycling since the operation of the Betim Recycling Plant in 2018 to 2022 were analyzed. They found as results of this research that the municipality

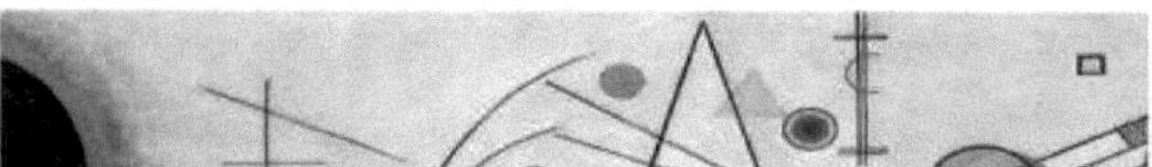

proposed the Municipal Plan of Integrated Management of Municipal Solid Waste, in 2015, according to current legislation, with goals that are being met, since 2018. Thus, in the last 5 years, it has recycled approximately 150,000 tons, transforming the recycled aggregate into rural road coverage, enforcing compliance with agenda 2030, preserving the ecosystem, and becoming a model city for other municipalities.

KEYWORDS: Sustainability, Agenda 2030, Environmental Responsibility, Municipal Plan for Integrated Solid Waste Management.

1. Introdução

Devido ao crescimento urbanístico, a indústria da Construção Civil tornou-se uma das mais importantes atividades socioeconómicas do país e uma das principais geradorasde resíduos sólidos nos municípios do Brasil, sendo um dos ramos de atividades que mais degrada o ambiente.

Em 2010 o Brasil gerava em torno de 33 000 000 toneladas de RSCD ao ano, ABRELPE, (2020) em 2021 passou a gerar cerca de 46 000 000, ABRELPE (2021). O aumento progressivo da geração de resíduos sólidos de construção e demolição (RSCD) produzidos no país motiva uma série de ações por parte do governo federal, dentre elas, podemos citar o desenvolvimento da Resolução Conama 307 de 05 de julho e de 2002, que estabeleceu diretrizes, normas e procedimentos para a gestão de resíduos da construçãocivil nos municípios, BRASIL (2002).

O presente estudo foi realizado no município de Betim (MG), pertencente à Região Metropolitana de Belo Horizonte Minas Gerais, possui uma área de 345,99 Km² e população de aproximadamente 450 000 pessoas, IBGE (2021).

Em 2015, a Secretaria do Meio Ambiente do município, elaborou o

Plano Municipal de Gestão Integrada de Resíduos Sólidos de Betim (PMGIRSB), obedecendo às diretrizes da Resolução Conama 307 de 05 de julho e de 2002, e a partir do PMGIRS, começou a ser implementada ações para reduzir os impactos causados ao meio ambiente derivados dos resíduos gerados pelo setor da construção e demolição, PMGIRSB (2015).

Segundo o PMGIRSB (2015) e, de acordo com dados coletados em 2009, o manejode resíduos sólidos de construção e demolição (RSCD) no município, assim como a maioria das cidades brasileiras, não vinha propiciando um manejo adequado aos Resíduos de Construção e Demolição/Resíduo de Construção Civil (RSCD/RCC), assim como não vinha seguindo a resolução do CONAMA n.º 307 e das normas da ABNT.

Em 2015 o município, através da secretária de Meio Ambiente, implantou 14Unidades de Recolhimento de Pequeno Volumes (URPV), sendo que cada URPV possui em médias três caçambas com capacidade volumétrica de 8m³, estima-se que cada URPV gere cerca de 3.000 toneladas de entulho/mês. Os resíduos eram recolhidos e enviados a Área de Triagem e Transbordo - ATT da Usina de Reciclagem de Resíduos de Construção e Demolição/Resíduo de Construção Civil, PMGIRSB (2015).

O PMGIRSB estabeleceu metas de reutilização e reciclagem dos RCD em 2015, para serem alcançadas em 2020, além disso, uma das metas do plano do município era de criar legislações municipais para melhor gestão e, em 2018 foi criada a LEI N.º 6412, DE 12 DE SETEMBRO DE 2018 que dispõe sobre a Política Municipal de Saneamento Básico, Prefeitura Municipal de Betim (2018)

A gestão das políticas públicas do município é realizada pela ECOS - Empresa de Construção, Obras, Serviços, Projetos, Transporte e Trânsito de Betim, a empresa foi transformada em autarquia municipal em 2018 através da lei 6381/2018, Câmara Municipal De Betim (2018).

O presente estudo visa contribuir com discussões teóricas e práticas

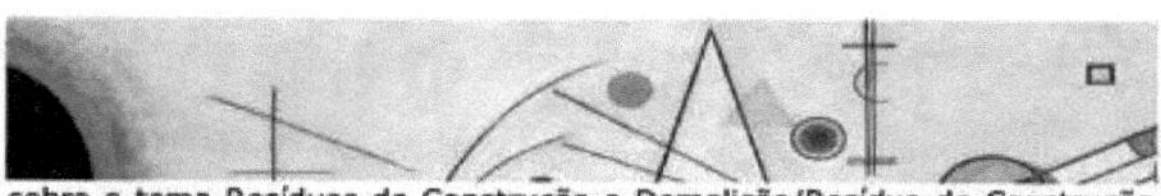

sobre o tema Resíduos de Construção e Demolição/Resíduo de Construção Civil no meio científico além de estimular os gestores municipais a praticar a gestão dos RCD conforme resolução do CONAMA 307, ABNT entre outras normativas para preservar o meio ambiente, avaliando a gestão dos RCD da cidade de Betim nos períodos de 2015 a 2022.

Em consideração a isso, este artigo teve por objetivo diagnosticar a gestão dos RSCD, em Betim.

2. Metodologia

Este trabalho, utilizou o método de pesquisa aplicada, de natureza descritiva, a partir de uma abordagem qualitativa, por meio de um estudo de caso.

O cenário de estudo foi o município de Betim - MG, localizado a 30 km de Belo Horizonte, região metropolitana de Minas Gerais - um dos municípios da região sudeste com maior crescimento urbanístico conforme o IBGE, (2020).

Realizou-se uma leitura do Plano de Municipal de Gestão Integrada de Resíduos Sólidos de Betim para entendimento da situação da gestão e das propostas elencadas, após a leitura realizada, foi efetuada a coleta dos dados da geração de resíduos do ano de 2015 do PMGIRSB e das metas descritas no documento para o trato dos RSDC.

Conforme Marconi e Lakatos (2017), o método quantitativo-descritivos tem como finalidade o delineamento ou a análise das características de fatos ou fenômenos, a avaliação de programas ou o isolamento de variáveis principais ou chave.

Neste contexto, foi realizada a coleta dos RSCD gerados, no período de 2018 a 2022 – anual, mensal e diário- com gestores da empresa ECOs – Empresa de Construções, Obras, Serviços, Projetos, Transporte e Trânsito

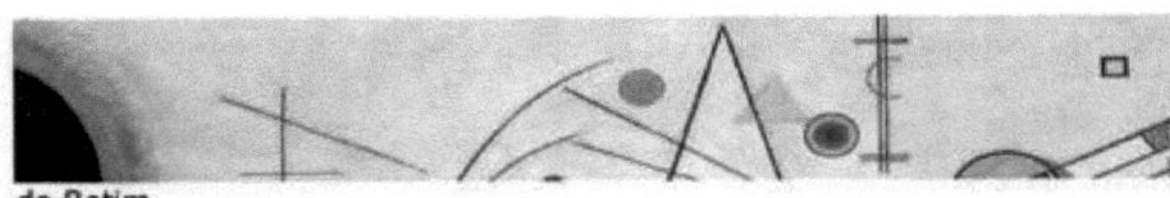

de Betim.

Foi realizado um estudo de amostragem da quantidade de Resíduos da Construção Civil geradas no município de Betim compreendidos entre os anos 2015 e 2022 e da geração de RSCD e reciclagem desses resíduos entre 2018 e 2022, após esse levantamento, foram elaborados os gráficos para melhor compreensão dos dados.

Os dados foram tratados utilizando recursos da estatística descritiva, a saber, média e porcentagem, para posterior comparação entre os resíduos gerados, no período de 2015 e 2022.

3. Resultados e Discussão

De acordo com o Panorama da Abrelpe (2021), em 2021 houve um aumento dos Resíduos Sólidos de Construção e Demolição (RSCD) gerados no Brasil de 5,5% em relação ao ano anterior. Em todo o país, foram recolhidos cerca de 47 milhões de toneladas por ano, sendo que o Sudeste tem percentual de 34 por cento desse total, gerando cerca de 24 milhões de resíduos por ano.

O município de Betim possui uma das maiores taxas de crescimento do Estado de Minas Gerais, e apresenta números expressivos de resíduos sólidos sendo que, de acordo com o Plano Municipal de Gestão Integrada de Resíduos Sólidos de Betim - PMGIRSB, em 2014 foram gerados um total de 82.152.120 kg/ano enquanto que Resíduos Sólidos da Construção, demolição e reforma representam aproximadamente 45% desse total dos resíduos gerados no município (PMGIRSB, 2015).

A Prefeitura Municipal de Betim, para colocar a usina em funcionamento, precisou adquirir novos equipamentos, a troca do maquinário foi necessária, já que, apesar de ter sida montada em 2011, a usina nunca havia funcionado até 2018. Ainda segundo a prefeitura, ela

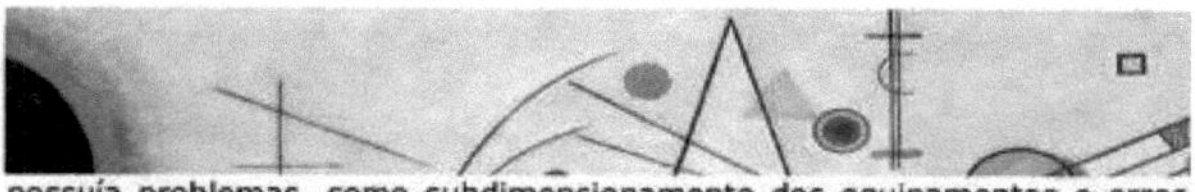

possuía problemas, como subdimensionamento dos equipamentos e erros emsua montagem, ECOs (2018).

Segundo o Plano Municipal de Gestão Integrada de Resíduos Sólidos de Betim (PMGIRSB), em 2015, período em que ainda não havia prática de reciclagem no município, foram geradas cerca de 103 toneladas por dia de RSCD, sendo que em 2022, a usina passou a receber uma média de 150 toneladas por dia para serem reciclados.

Figura 1: Controle de produção diária da usina referente aos meses de janeiro a junho de 2022.

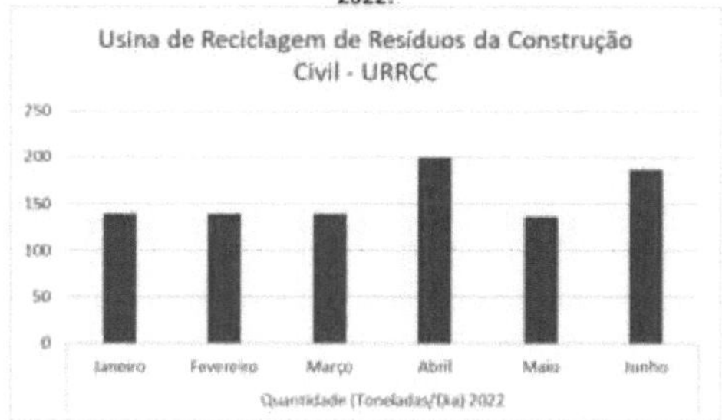

O cálculo da média, considerou 20 dias de produção mensal devido às paradas porocasião de manutenções preventivas e corretivas, assim como eventos sazonais.

No levantamento de janeiro a julho de 2022, a média de demanda foi de 150 toneladas/dia considerando 20 dias/mês de trabalhos na usina, totalizando aproximadamente 3000 toneladas/ mês. Com esta média mensal em 12 meses teremos 36 000 toneladas geradas na cidade de Betim - MG.

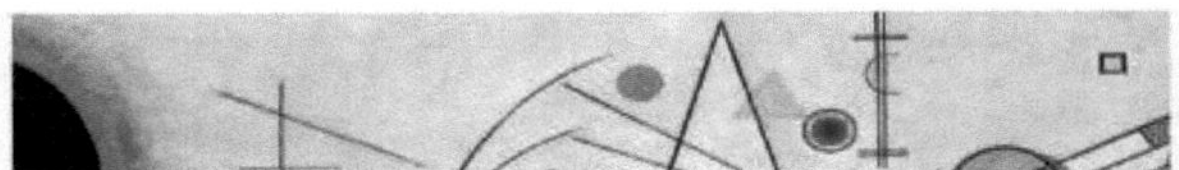

Figura 2: Controle de produção mensal da usina referente aos meses de janeiro a junho de 2022.

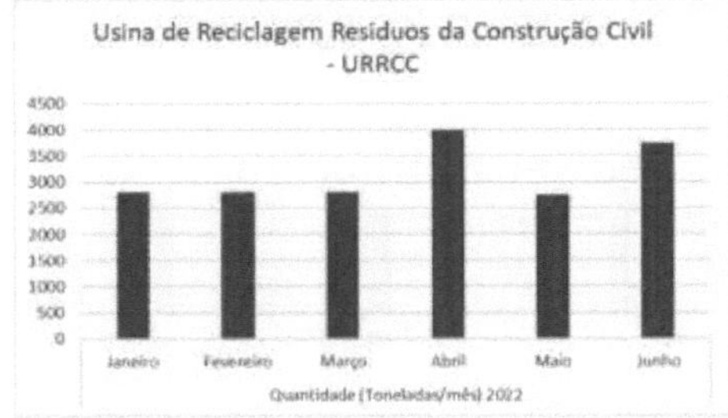

A prefeitura de Betim, através da Empresa ECOs, desde 2018 realiza a britagem dos RSCD gerados no município e começou com capacidade de reciclar 2.500 toneladas e em 2021, estava realizando a britagem média de 2800 toneladas/mês.

A usina possui capacidade máxima de britagem com potencial de aumento da produção, assim como do recolhimento dos RSCDs.

Figura 3: Controle de produção anual da usina desde a sua inauguração em 2018 até 2022

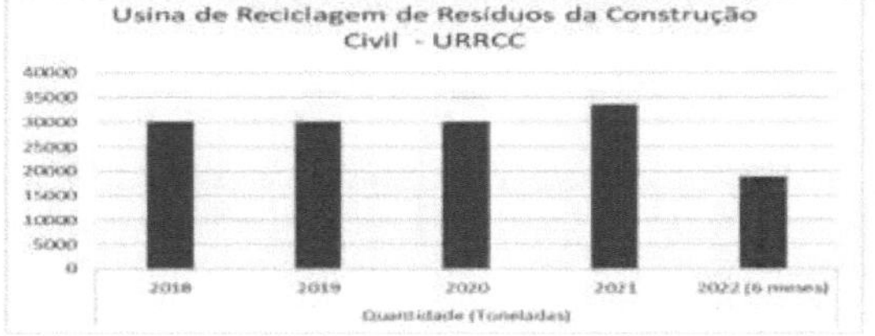

2435

Segundo o DSA/ECOs Usina de Reciclagem de Resíduos da Construção Civil - URRCC da cidade de Betim - MG em 2018, reciclou 30.000 toneladas por ano na sua capacidade máxima e em 2022 atingiu 33.000 toneladas por ano, com previsão de aumento de acordo com a demanda aplicada.

Os RCD diários gerados entre 2015 e 2022 representam um percentual de aumento de 45,63 por cento num intervalo de apenas 7 anos, com uma média de 6,5 por cento de aumento anual como vemos na figura 4.

Figura 4: RCD gerados entre 2015 e 2022 na cidade de Betim-MG

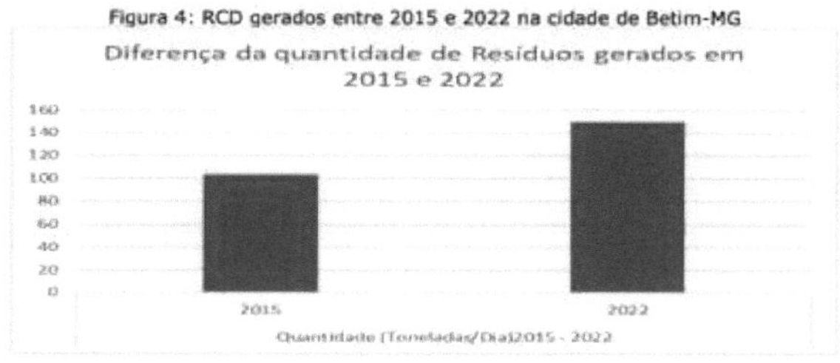

Se prevermos uma média percentual de 2022 a 2030, com o ramo da construção mantendo a escala de produção dos últimos 7 anos, teremos um aumento de 52% de RSDCna cidade de Betim, o que indica uma necessidade de manter a usina em funcionamento para evitar que esses resíduos sejam destinados de forma incorreta, causando danos aomeio ambiente.

A cidade de Betim se destaca na gestão dos resíduos sólidos fazendo cumprir as metas do PGRS, desde 2015 a prefeitura já retirou de possíveis aterros clandestinos e "botaforas", em média 250.000 toneladas de resíduos sólidos no período de 7 anos. Até 2030, podemos estimar o dobro desse valor, ou seja, serão no mínimo 520.000 toneladas dispostos de forma

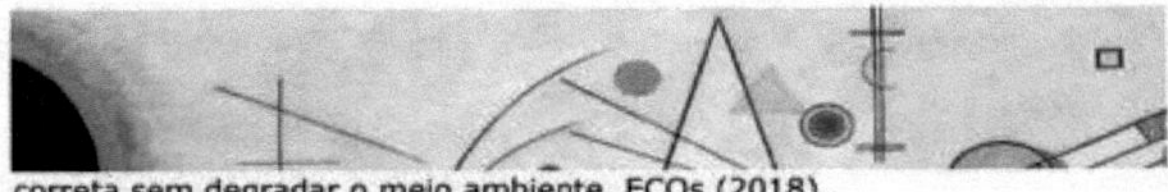
correta sem degradar o meio ambiente, ECOs (2018).

Enfatiza-se que a construção civil é um dos setores que mais degrada o meio ambiente, impactando a biodiversidade, poluindo a atmosfera, reduzindo as reservas hídricas, e principalmente gerando resíduos, especialmente quando a obra não é feita com planejamento adequado (LARUCCIA, 2014).

Para evitar a degradação do meio ambiente frente ao descarte incorreto dos RSCD, geração destes resíduos, atendendo às necessidades de consumo, sem comprometer o meio ambiente e as futuras gerações, obedecendo as legislações e cumprir as propostas daagenda 2030 do Brasil.

Conforme Gnecco (2018), novas tecnologias vêm avançando para minimizar tais impactos ambientais como a tecnologia BIM em inglês *Building Information Modeling*, porém, a ferramenta ainda se posiciona como uma tecnologia com um preço alto, o software BIM custa em média R$1400,00/Mês (Hum mil e quatrocentos reais) cada licença, além da necessidade de contratar uma consultoria para implantação e utilização, logo, as pequenas empresas e proprietários de imóveis - que vão demolir, construir ou reformar - ainda não tem acesso a tecnologias como o BIM e esse público representa uma grande parcela de geradores de resíduos, Autodesk (2022).

A ferramenta é uma plataforma tecnológica aplicada a construção civil que possibilita modelar os projetos e detalhes das especificações das instalações ou da edificação evitando desperdícios e custos na construção. (CATELANI, 2016).

Marinho (2017) propõe que: o BIM é capaz de gerenciar e armazenar informaçõese dados, durante todo o ciclo de vida da obra, desde a concepção e a conceituação da ideia, desenvolvimento do projeto, a construção, e também após a obra pronta, na sua fase de utilização

Brito e Ferreira (2015) define BIM como uma metodologia de projeto

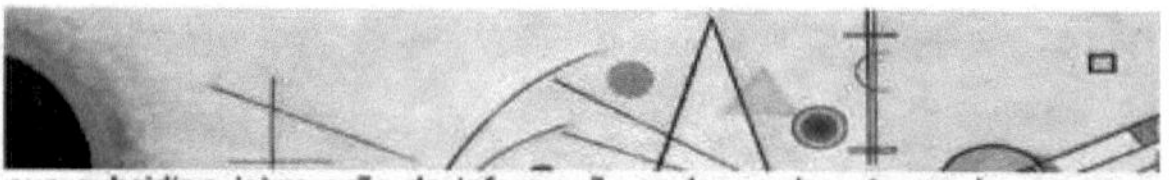

que subsidia a integração da informação ao longo das etapas de processo de projeto, gestão de projetos e obras.

Portanto, mais do que a pratica de reciclagem, é necessário estimular a redução de geração dos RSCD.

Dito isso, a proposta desse estudo, é demonstrar que o cumprimento da agenda 2030 Brasil é possível através do cumprimento das legislações, além de incentivar outros municípios a realizar a gestão correta dos resíduos sólidos da construção, efetuando o reaproveitamento do mesmo após reciclado, já que os materiais resultantes do processo de reciclagem podem ser usados na fabricação de tijolos, blocos, meio fios e outros, evitando assim a degradação do meio ambiente, deixando para as futuras gerações um ecossistema preservado garantindo-lhes uma qualidade de vida, além de propor a redução do RSCD através de pesquisas de novas tecnologias.

4. Considerações Finais

A pesquisa realizou o diagnóstico da gestão dos RSCD da cidade de Betim com dados anuais, mensais e diários desde a elaboração do PMGIRSB que ocorreu em 2015, trouxe ainda conceitos sobre Resíduos de Construção e Demolição/Resíduo de Construção Civil em conformidade com a resolução do Conama 307 de 05 de julho de 2002 no meio científico além de estimular os gestores municipais a praticar a gestão dos RCD conforme resolução do CONAMA 307, ABNT entre outras normativas para preservar o meio ambiente, avaliando a gestão dos RCD da cidade de Betim nos períodos de 2015 a 2022.

O município de Betim segue as normas para cumprir as suas metas de reciclagem dos resíduos sólidos da construção demolição, a cidade trabalhou para reduzir os impactos ambientais.

Com o objetivo ambiental, a quantidade de RSCD que teriam um

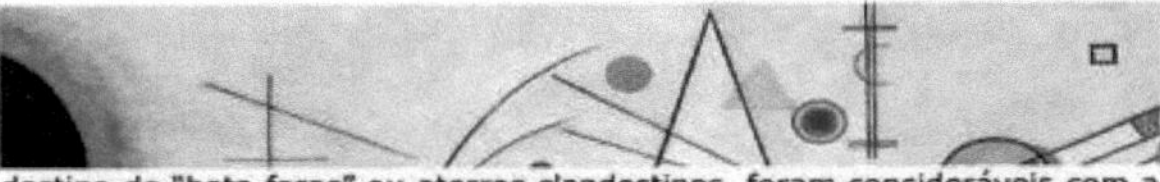

destino de "bota foras" ou aterros clandestinos, foram consideráveis com a criação da usina.

Para evitar esses "Bota foras", descartes clandestinos que degrada o meio ambiente, faz-se necessário estudos especiais de gerenciamento de RSCD assim como o fomento da reciclagem, atendendo as necessidades de consumo sem prejudicar o meio ambiente, cumprindo as legislações e fazendo cumprir a agenda 2030 Brasil.

A Reciclagem dos RSCD é um passo importante para a conquista de um ecossistema sustentável, e em parceria dos poderes públicos e privados, o trato dos resíduos se torna mais acessível, no caso apresentado, foi de extrema importância a criação de uma autarquia no município para desburocratizar a gestão pública.

Atender ao CONAMA/307 realizando o Plano municipal de gestão integrada de resíduos sólidos é o primeiro passo para a eficácia da gestão dos RSCD e o caminho para garantir o cumprimento da agenda 2030, conforme demonstrado na pesquisa a iniciativa de realizar o PMGIRSB foi o que levou o município concretizar as metas estabelecidas.

Faz-se necessário pesquisa de novas tecnologias para evitar a geração dos resíduos sólidos da construção e demolição.

Incentivar outros municípios reciclar através de exemplos como a cidade de Betim - MG, é importante para desburocratizar a ação e demonstrar que é possível o Brasil chegar em 2030 como o país que conseguiu cumprir a sua agenda.

Por fim, estimular a adoção de técnicas de reciclagem e a boa prática de gestão dos RSCD para a preservação do meio ambiente, é garantir uma vida melhor a geração atual e as futuras gerações.

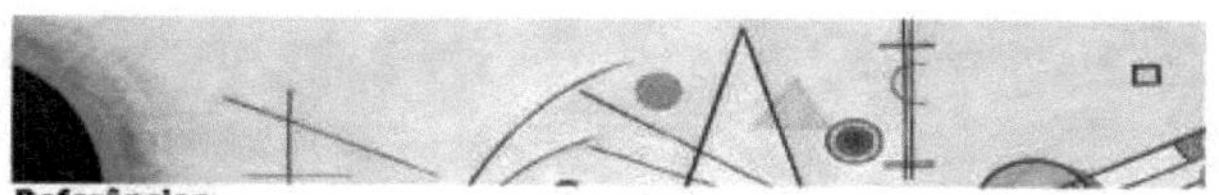

Referências

ABRELPE - Associação Brasileira de Empresas de Limpeza Pública e Resíduos Especiais In: **Panorama 2021**. [S. l.], 2021. Disponível em: https://abrelpe.org.br/panorama/. Acesso em: 8 jan. 2022.

AUTODESK (EUA). Revit: software BIM para projetistas, construtores e desenvolvedores. *In*: **Revit: software BIM para projetistas, construtores e desenvolvedores**. 1.0-RC2. [*S. l.*], 2022. Disponível em: https://www.autodesk.com.br/products/revit/overview?term=3-YEAR&tab=subscription. Acesso em: 25 jul. 2022.

BRASIL - CÂMARA DOS DEPUTADOS. Ministério das relações exteriores, divisão do meio Ambiente, com a colaboração do Programa das Nações Unidas para o meio ambiente, PNUMA. Ação Parlamentar, 56, 1995. 56. **Conferência das nações unidas sobre o meio ambiente e desenvolvimento**, Brasília-DF: Centro de Documentação e Informação Coordenação de Publicações, v. 1, p. 472, 1995.

BRASIL - Secretaria Nacional de Saneamento Ambiental. Rede de capacitação e extensão tecnológica em saneamento ambiental. ReCesa, 2007. **Resíduos sólidos: PGRSU -Plano de gestão de resíduos Sólidos urbanos**: : guia do profissional em treinamento, Belo Horizonte- MG: Artes Gráficas Formato LTDA, v. 1, p. 100, 2007.

BRITO, D. M. de; FERREIRA, E. de A. M. Avaliação de estratégias para representação e análise do planejamento e controle de obras utilizando modelos BIM 4D. **Ambiente Construído**, Porto Alegre, v. 15, n. 4, p. 203-223, out./dez. 2015.DOI: http://dx.doi.org/10.1590/s1678- 86212015000400047.

CÂMARA MUNICIPAL DE BETIM. Lei Ordinária nº 6381, de 29 de junho de 2018. DISPÕE SOBRE A TRANSFORMAÇÃO DA EMPRESA DE CONSTRUÇÕES, OBRAS, SERVIÇOS, PROJETOS, TRANSPORTES E TRÂNSITO DE BETIM - ECOS EM AUTARQUIA PÚBLICA MUNICIPAL, E DÁ OUTRAS PROVIDÊNCIAS. **Lei Ordinária Nº 6.381/2018**, Betim, 2018

CATELANI, Wilton Silva (Brasil). Cbic - Câmara Brasileira da Indústria da Construção.

Coletânea Implementação do BIM Para Construtoras e Incorporadoras: Fundamentos BIM - Acesso em: 12 abr. 2017.

ECOs - **Empresa de Construções, Obras, Serviços, Projetos, Transporte e Trânsito Informativo anual, 2019.** Prefeitura Municipal de Betim (Minas Gerais). Betim -MG, 2019.

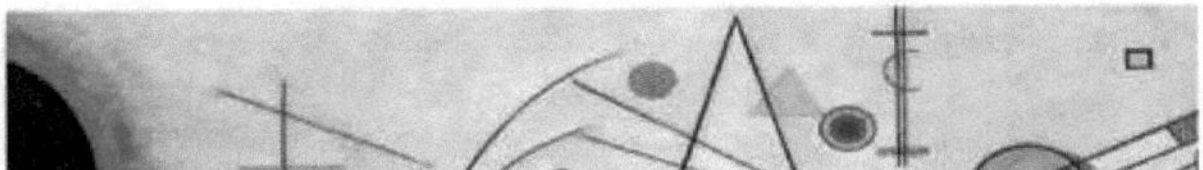

- **Empresa de Construções, Obras, Serviços, Projetos, Transporte e Trânsito Informativo anual, 2020.** Prefeitura Municipal de Betim (Minas Gerais). Betim -MG, 2020.

- **Empresa de Construções, Obras, Serviços, Projetos, Transporte e Trânsito Informativo anual, 2022.** Prefeitura Municipal de Betim (Minas Gerais). Betim -MG, 2022.

FRASSON, S.. **Usinas de reciclagem de entulho como agentes na valoração dos resíduos gerados pela construção civil**, [s. L.], 12017.

GNECCO, verônica. **Bim para gestão de resíduos na construção civil: estudo de casona obra do cras do município de biguaçu/sc**. 2019. 150 f. TCC (Bacharelado) - Centro Tecnológico da Universidade Federal de Santa Catarina, Santa Catarina, 2018.

IBGE - Instituto Brasileiro de Geografia e Estatística. População 1980. In: Governo Federal. IBGE. População. [S. l.], 2021. **População total - 1980 - 2010**. Disponível em: https://brasilemsintese.ibge.gov.br/populacao/populacao-total-1980-2010.html. Acesso em: 1 Maio. 2022.

LARUCCIA, Mauro. Sustentabilidade e Impactos Ambientais da Construção Civil. **Artigo**,[s. *l.*], 2014.

MARCONI, Marina d; LAKATOS, Eva. **Fundamentos de metodologia científica**. 8. ed. atual. São Paulo: Atlas, 2017. ISBN 978-85-970-1076-3.

MEU Município. *In*: **Dados de Betim**. 1.0-RC2. [*S. l.*], 2015. Disponível em: https://meumunícipio.org.br/perfil-municipio/3106705-Betim-MG?exercicio=2015. Acesso em: 14 jun. 2022.

PMGIRSB - **Plano municipal de gestão integrada de resíduos sólidos de Betim**. PMB - Prefeitura Municipal de Betim (Minas Gerais). 2015., Betim -MG, p. 1-128, 2015.

PREFEITURA MUNICIPAL DE BETIM. **Lei nº 6412, de 12 de setembro de 2018**. DISPÕE SOBRE A POLÍTICA MUNICIPAL DE SANEAMENTO BÁSICO E DÁ OUTRAS

PROVIDÊNCIAS. Betim Minas Gerais, 12 set. 2018.

SEMMAD - **Secretária municipal de meio ambiente e desenvolvimento sustentável**.PMB - Prefeitura Municipal de Betim (Minas Gerais). 2015., Betim -MG, p. 1-128, 2015.

Printed by Books on Demand GmbH, Norderstedt / Germany